Determinants and Stratification of Household Carbon Footprint

Sunil Nautiyal • Mrinalini Goswami
A. Premkumar

Determinants and Stratification of Household Carbon Footprint

Understanding the Prospect for Climate Change Mitigation and Socioeconomic Co-benefits

Springer

Sunil Nautiyal
Centre for Ecological Economics
and Natural Resources
Institute for Social and Economic Change
Bengaluru, Karnataka, India

Mrinalini Goswami
Centre for Ecological Economics
and Natural Resources
Institute for Social and Economic Change
Bengaluru, Karnataka, India

A. Premkumar
Centre for Ecological Economics
and Natural Resources
Institute for Social and Economic Change
Bengaluru, Karnataka, India

ISBN 978-981-96-2738-7 ISBN 978-981-96-2739-4 (eBook)
https://doi.org/10.1007/978-981-96-2739-4

This Springer imprint is published by the registered company Springer Nature Singapore Pte Ltd.
The registered company address is: 152 Beach Road, #21-01/04 Gateway East, Singapore 189721, Singapore

Foreword

As we stand at a critical juncture in our global fight against climate change, I am pleased to present this comprehensive analysis of household carbon footprint (HCF), a study that bridges the gap between environmental sustainability and socioeconomic development. This research comes at a time when understanding the intricate relationship between human activity and environmental impact may hold the key to shape our climate action responses.

We know that the impact of climate change is a global concern, but at the same time it also manifests as a deeply personal issue affecting millions of Indian households in diverse ways. As we strive to balance economic growth with sustainability, understanding the carbon footprint of our homes becomes essential. This book, *Determinants and Stratification of Household Carbon Footprint: Understanding the Prospect for Climate Change Mitigation and Socioeconomic Co-benefits*, offers an extensive analysis of household emissions throughout India, in addition to providing critical insights into the carbon emission contributions of various socioeconomic groups and explores potential policy interventions to effectively address the resultant issues.

One of the defining strengths of this study lies in its comprehensive coverage of India's heterogeneous landscapes panning urban, peri-urban, and rural areas across multiple agro-climatic zones. By capturing regional variations in consumption behavior, energy use, and transportation choices, this research presents an unparalleled, detailed estimation of emissions. The findings emphasize how affluence and accessibility to modern technology drive emissions in high-income groups, while low-income households often have a lower carbon footprint yet remain the most vulnerable to climate impacts.

In the Indian context, where over 65% of the population still resides in rural areas and economic disparities are stark, the findings of the study underscore the urgent need for targeted interventions. The study elucidates how increasing urbanization is altering consumption patterns, thereby contributing to rising emissions from food, energy, and transport sectors. Concurrently, it highlights sustainable practices that already exist in many rural communities that can be strengthened and adapted for larger-scale climate mitigation efforts.

The varying patterns of carbon emissions across different socioeconomic strata and geographical locations within India, one of the most populous and rapidly developing nations, are particularly pertinent and relevant to contemporary climate change discourse. This is because as nations worldwide address their commitments to reduce greenhouse gas emissions, the household as a fundamental unit of emission generation becomes important. Hence the relevance of this study to contemporary climate change discourse is significant.

This research is particularly compelling due to its holistic approach in examining household carbon footprints. The study meticulously analyzes various aspects of daily life—from energy consumption and transportation to food habits and consumption patterns. The findings indicate that household energy emissions constitute a significant 39.22% of total emissions, while transportation accounts for approximately 19.73%. These results are of considerable importance as India projects a six to thirteen-fold increase in residential electricity consumption by 2047, according to projections of the NITI Aayog.

The timing of the study is also relevant because India has recently achieved near-universal household electrification, with 750 million people gaining access to electricity between 2000 and 2019. This massive transformation in energy access presents both challenges and opportunities for sustainable development. The research provides critical baseline data that can assist policymakers in identifying specific intervention points for emission reduction while ensuring that development goals are not compromised, which will hopefully result in policies that are both environmentally conscious and socially inclusive.

Perhaps, the most significant contribution of this study lies in its potential to inform targeted policy interventions that can simultaneously address climate change and poverty alleviation. Through the analysis of carbon footprint patterns across different socioeconomic strata, it becomes possible to formulate policies that promote low-carbon development pathways while improving the quality of life for underprivileged communities. The research underscores the feasibility of integrating sustainable practices into development programs, thereby creating a synergetic approach that benefits environmental conservation and social progress.

The findings pertaining to food-related emissions demonstrate how traditional Indian dietary patterns result in significantly lower carbon footprints as compared to diets that rely more on meat or dairy products. This knowledge can be instrumental in promoting sustainable dietary choices while preserving cultural practices that are not only environment-friendly but inherently healthier.

The rigorous methodology and extensive primary survey across different socio-economic backgrounds make this study a valuable reference for researchers, development practitioners, and government agencies working toward a sustainable future. I commend the researchers for this comprehensive analysis and believe that this book will serve as a crucial reference point for policymakers, researchers, and practitioners working at the intersection of climate change mitigation and sustainable development.

I extend my deepest appreciation to the authors and the research team for their meticulous and painstaking efforts in bringing out this work, which is more than just a compilation of data and offers a road map for balancing India's developmental goals with its environmental responsibilities. Let us use this knowledge as a stepping stone to a more sustainable and equitable future.

National Commission for Women (NCW), Govt. of India
New Delhi, India
Forest Force, Govt. of Karnataka, Bengaluru, India
Meenakshi Negi

Preface

The study on carbon footprints of Indian households was conducted by the Institute for Social and Economic Change (ISEC), Bengaluru, as a part of a broader project "Supporting the Institutionalisation of Capacities on Climate Change Studies and Actions," implemented by the Ministry of Environment, Forest and Climate Change (MoEFCC), Govt. of India, and Deutsche Gesellschaft für Internationale Zusammenarbeit (GIZ). The study necessitates evaluation of household carbon emissions to understand the extent of nation-wide emissions and variations across socioeconomic strata and agro-ecological zones. The methodological design adopted for this study was cross-sectional with consultative approach. The primary objective of the study was to determine the carbon emissions of households and their changing lifestyle followed by emission comparisons among 13 locations in nine agro-ecological zones and across levels of urbanizations (rural, peri-urban, and urban). The carbon footprint calculations were based on primary data on consumptions by households collected through household questionnaire survey of more than 2100 households with uniform coverage of agro-ecological zones, level of development, and socioeconomic conditions. The survey tool used was an extensive questionnaire with questions under broad categories of socioeconomic description of households, income and expenditure categories, land resources and assets, consumptions with subdivisions including energy, food, travel, and transport and all other non-durables, behaviour and determinants of consumption along with associated regulations, and barriers for sustainable consumption.

The carbon footprint of Indian households is estimated around 5265 $KgCO_2e$ per household per year with a corresponding per capita carbon footprint of 1273 $KgCO_2e$ per year. While only basic household carbon footprint (HCF) is calculated excluding the items that are non-prevalent among the lower income groups, the average HCF was calculated to be lower than the average. The impact of economic well-being on carbon emission is evident from the upward trend of increasing footprint with respect to increasing household's annual income. Higher income group has five times higher emission than the lower income group. Further assessment of carbon footprints from different sectors (food, energy, and travel) has revealed variations among socioeconomic classes and across geographical locations. Inconsistencies in consumption among households and rural-urban divide have emerged very prominently in the study results. Respondents mostly believe that the environmental concerns are overstated. Motivation to shift towards low-carbon alternatives or reduction of HCF is largely guided by well-being of the family, mainly, reduction of health risks due to climate change. The majority of the surveyed households may consider adoption of low-carbon alternatives if they don't have to compromise with cost and quality of the existing item or practice. Government policies and subsidies have been given equal importance by the rural population. However, the affluent section of the society stated that increased ease of access to available technology/alternatives and awareness should be primary strategies for HCF reduction. Along with deepening sustainable lifestyles, focus should be on judicious utilization of India's tradable carbon surplus for the economic well-being of disadvantaged populations. The changing consumption pattern of mounting middle-income group and high-income group are to be guarded through policy innovations for enhancing knowledge, motivation and access to low-carbon alternatives; eco-labelling, taxation, penalty, and subsidies. Popularizing traditional sustainable lifestyles and the concept of self-reliant landscapes, and production system through circular economy should be aggressively taken up for household emission reduction. Compensating the low consumption lifestyle irrespective of rural-urban divide can be effective in maintaining a low average HCF of Indian households. It is expected to contribute towards awareness building, knowledge exchange, strengthening capacities for enhanced science-policy-practice and more informed bottom-up strategy formulation for low-carbon lifestyle with subsequent socioeconomic development. The following highlights of this study describe a holistic perspective of the HCF in India and the potential of India's tradable carbon surplus which can be judiciously utilized for the economic well-being of disadvantaged populations.

In-depth analysis was done on collecting information from empirical field studies on: quantification of carbon footprint of Indian households; consumption of energy/resources with an increase in household income; pathways for sustainable lifestyles including those in traditional lifestyles; the impact of COVID-19 led lockdown on consumption behaviour of the Indian households.

The objectives also include ground-truthing of sectoral emissions of households and providing concrete suggestions to stakeholders to mainstream mitigation actions identified on the basis of sample survey in target sectors. This comprehensive bottom-up analysis will identify the sectoral differences in HCF and identify potential pathways to HCF reduction.

This will aid the policy-makers to take informed science-based actions in these sectors. The authentic empirical data set is also expected to provide a direction to adopt required changes in the frameworks of national level data (e.g. NSSO) collection.

While the economic liberalization initiated in 1991 and programmes addressing energy poverty in India have lifted many Indian households from absolute poverty and have created millions of middle-class families (mostly centred in cities) who demand more energy for their sustenance. In fact, the growing middle class, with up to 300 million people, implies an acute shift in the consumption amount and type goods, and services which is further triggering the increase of the embodied emissions. The contribution of household carbon is significant which is primarily guided by income and consumer behaviour. Variation in household emissions in rural and urban India among different socioeconomic classes is also notable. Studies also show that higher emissions do not necessarily lead to better well-being in households. In contrast to developed countries, developing countries have the opposite pattern of emissions with degree of urbanization, where households in urban areas emit more.

A comprehensive understanding of emissions patterns across different groups and understanding on sectoral emissions is needed to identify emissions reduction pathways that can contribute to mitigation planning through both policy and awareness building. Another important aspect in implementing mitigation strategies is the acceptance of mitigation strategies by different social and economic sectors considering their development goals and different social and cultural norms.

While mitigation can be addressed from a top-down perspective through increased investments on a large scale for renewables, the role of households as end-users and demand-side mitigation often remained unaddressed in Indian context. This study focuses on the issue of mitigation at a national level solely from a demand-side perspective. It encompasses all the sectors that are covered in India's NDCs since household's consumption behaviour dictates the need for a given good/service in the market. This, on one hand, impacts the GDP of the country and, on the other hand, affects the respective sectoral emissions. For instance, a given household cannot have a direct influence on the type of electricity (i.e., coal powered or renewable) it receives but can decrease its HCF by minimizing electricity consumption through increased awareness and/or through installing energy-efficient appliances. The planned primary survey addresses this aspect of Indian households and identifies pathways for sustainable lifestyles.

We extend our sincere thanks to the GIZ, New Delhi, for the funding support to conduct the study. We would like to thank Ms Meghana Kshirsagar, Mr Shailendra Dwivedi, and team from GIZ for their constructive suggestions and support throughout the study. We extend our sincere thanks to Dr. Ramana Gudipudi (Potsdam Institute for Climate Impact Research, Germany) for proposing the collaboration and initiation of the work with his valuable insights. We are deeply thankful to all the stakeholders across 11 agro-climatic zones and all the respondents who took part in the survey. We whole heartedly thank all the field staffs for their support, valuable inputs, immense help, and cooperation during the entire period of study. We acknowledge with warm thanks the help which we have received directly or indirectly from the people of the study region and thankful for expressing their views in the context of the current study. We also extend our sincere thanks to Sacred Heart College, Tirupattur, and Madhya Pradesh Vigyan Sabha for their support in organizing the Focus Group Discussions in Tamil Nadu and Madhya Pradesh. We would like to extend our sincere gratitude to the Director, Institute for Social and Economic Change (ISEC), Bengaluru, for providing all the facilities and support required to conduct this study. We are grateful to the G.B. Pant National Institute of Himalayan Environment (NIHE), for its support in providing access to the facilities. We would like to extend our thanks and appreciation to Ms. Nupur

Verma, Associate Editor – Books, and Ms. Nandhini Viswanathan, Production Editor, for their kind help, support, and assistance throughout the entire publication process. Last but not least, we would like to extend our thanks and appreciation to the research scholars at CEENR (ISEC) for being so helpful.

Bengaluru, Karnataka, India

Sunil Nautiyal
Mrinalini Goswami
A. Premkumar

Contents

About the Authors

Sunil Nautiyal (AvH-Fellow) is a Professor at the Centre for Ecological Economics and Natural Resources (CEENR), Institute for Social and Economic Change (ISEC), Bengaluru and Honorary Professor at the University of Ladakh. Presently, Dr. Nautiyal is serving as Director, G.B. Pant National Institute of Himalayan Environment (NIHE), Almora, Uttarakhand, India. His specialization includes areas of Natural Resource Management and Conservation, Protected Area Management and HWC&C GIS and Remote Sensing for Landscape Research, Urban-rural interface and Sustainability, Climate Change, Socio-ecological Development (Socio-Ecologist). Dr. Nautiyal has published over 200 scientific papers/articles and 14 books/monographs. Dr. Nautiyal is a recipient of several national/international prestigious fellowships/awards. He is on the Editorial/Reviewer Board of several international/national journals. Dr. Nautiyal has been working in close coordination with scientists in India, Germany, UK, Australia, Japan, Bangladesh, and the USA. His research contribution has large policy implications in the area of socio-ecological development and biodiversity conservation.

Mrinalini Goswami is a researcher in ecological science and is associated with GB Pant National Institute of Himalayan Environment and Centre for Ecological Economics and Natural Resources, Institute for Social and Economic Change, Bengaluru. She received her Ph.D. in Environmental Science for her research on sustainability of ecosystem-based livelihoods in peri-urban landscapes. She has been engaged in research on varied topics, including environmental sustainability, climate change, natural resource management, water management, and urban development. Dr. Goswami has published scientific papers/articles and books in the field of socio-ecology, environment and ecology. Dr. Goswami also has substantial work experience in the development sector, which has motivated her towards interdisciplinary approach in environmental research.

A. Premkumar is currently working as an Assistant Professor at the Department of Economics, Kristu Jayanti College (Autonomous), Bengaluru, Karnataka, India. Prior to this, he worked on various research projects at the Centre for Ecological Economics and Natural Resources, Institute for Social and Economic Change (ISEC), Bangalore. He received his Ph.D. in Economics from Thiruvalluvar

University, Tamil Nadu. He specialized in Environmental Economics, mainly on Climate Change, Solid Waste Management, and Household Carbon Footprint. To his credit, he has written and published several articles in Scopus and UGC Care-listed journals. He has also presented many research papers at international and national conferences.

Global Climate Dialogues: India's Position and Perspectives

1

1.1 Introduction

On a global scale, climate change is marked by increasing temperatures, which result in more frequent and severe weather events such as hurricanes, droughts, and heat waves. Melting polar ice caps and glaciers are causing sea levels to rise, posing threats to coastal areas. Warmer ocean temperatures are leading to coral bleaching and disruptions in marine ecosystems. Shifts in precipitation patterns are impacting freshwater resources and agricultural yields, causing food and water shortages. Biodiversity is at risk as habitats are changed or destroyed. Climate change is worsening social and economic inequalities, especially for vulnerable groups. Global efforts to combat climate change include reducing greenhouse gas emissions and shifting to renewable energy sources. Adaptation strategies are crucial to manage the unavoidable effects of climate change. India's strategy towards addressing climate change has transformed over time, carefully managing its global obligations alongside national development goals. In international climate talks, India has consistently supported fairness principles and historical accountabilities (Ahmed 2023), progressively embracing more proactive approaches, highlighted by ambitious pledges at COP26 in Glasgow (Dubash et al. 2022). The nation has established ambitious targets for reducing emissions, such as reaching 450 gigawatts of renewable energy capacity and reducing carbon emissions by 33% by 2030 (Teske 2019).

1.1.1 Global Climate Change Conventions

In India, national and state action plans have been devised to tackle climate change, concentrating on constructing resilient infrastructure and enhancing resource planning (Sinha et al. 2018). Initially, these efforts were more motivated by international commitments than by domestic imperatives. The nation grapples with the challenge

S. Nautiyal et al., *Determinants and Stratification of Household Carbon Footprint*,
https://doi.org/10.1007/978-981-96-2739-4_1

of shifting to a low-carbon economy while addressing its developmental imperatives, especially evident in its urbanization strategies (Mukhopadhyay 2021).

This section offers an in-depth overview of the key international agreements and conventions aimed at addressing the urgent issue of climate change. These conventions, created through the joint efforts of various nations, concentrate on reducing the impacts of climate change and encouraging sustainable environmental practices. They signify essential progress in the global community's dedication to tackling environmental challenges and building a more resilient and sustainable future. Many nations have pledged to achieve net-zero greenhouse gas emissions by 2050, in line with scientific guidance to prevent the most severe effects of climate change.

1.1.1.1 United Nations Framework Convention on Climate Change (UNFCCC)

The United Nations Framework Convention on Climate Change (UNFCCC) was established in 1992 at the Earth Summit in Rio de Janeiro, Brazil. This pivotal convention laid the groundwork for global climate action by providing a structured approach for international negotiations aimed at addressing climate change. The key outcomes of the UNFCCC include the recognition of climate change as a major global challenge, the formulation of guiding principles for future climate initiatives, and the commitment of member countries to monitor and report on greenhouse gas emissions and climate policies. Additionally, the convention underscored the responsibility of developed nations to lead in emission reductions and pledged support to developing countries through financial and technological resources to assist them in meeting their climate objectives.

1.1.1.2 United Nations Convention to Combat Desertification (UNCCD)

Adopted in 1994 at a conference in Paris, France, the United Nations Convention to Combat Desertification (UNCCD) aims to tackle desertification and land degradation, especially in arid, semi-arid, and dry sub-humid regions. The convention underscores the importance of sustainable land management to restore and sustain land productivity. The key outcomes include the creation of national action plans to address desertification, the promotion of global cooperation and partnerships to support affected nations, and the inclusion of local communities in land management efforts. Additionally, the UNCCD focuses on mobilizing financial resources and technical assistance to effectively implement land restoration and sustainable development strategies.

1.1.1.3 Kyoto Protocol

Adopted in 1997 during the Third Conference of the Parties (COP3) held in Kyoto, Japan, the Kyoto Protocol was a landmark agreement in global climate policy. It established legally binding targets for developed nations to reduce their greenhouse gas emissions. The key outcomes of the protocol included the setting of specific emission reduction goals for industrialized countries, the introduction of flexible mechanisms such as emissions trading, the Clean Development Mechanism (CDM), and Joint Implementation (JI) to facilitate cost-effective compliance. Additionally, the protocol emphasized the principle of common but differentiated responsibilities, acknowledging that developed countries should lead in reducing emissions and provide support to developing nations through financial and technological means.

1.1.1.4 Rotterdam Convention

The Rotterdam Convention, established in 1998 at a conference in Rotterdam, the Netherlands, is an international agreement designed to regulate the Prior Informed Consent (PIC) process for hazardous chemicals and pesticides traded globally. The convention's key resolutions include the development of a global framework that requires exporters to disclose detailed information about hazardous substances and obtain approval from importing countries before shipment. Its main objectives are to enhance transparency, protect human health and the environment, and enable countries to make informed choices regarding the importation of dangerous chemicals.

1.1.1.5 Cartagena Protocol on Biosafety

The Cartagena Protocol on Biosafety was established in 2000 during a conference in Montreal, Canada. This international agreement focuses on ensuring the safe management, transport, and use of living modified organisms (LMOs) produced through modern biotechnology. The key outcomes of the protocol include the creation of a robust framework for assessing and managing the risks associated with LMOs, mandating that countries implement national biosafety regulations and perform risk assessments. The protocol also highlighted the need for informed consent, public involvement, and transparent information exchange between nations to safeguard biodiversity and human health from potential risks related to genetically modified organisms.

1.1.1.6 Stockholm Convention

Adopted in 2001 at a conference in Stockholm, Sweden, the Stockholm Convention is an international treaty designed to safeguard human health and the environment from Persistent Organic Pollutants (POPs), which are hazardous chemicals that remain in the environment and accumulate in the food chain. The key resolutions and highlights of the convention include the listing of specific POPs to be eliminated or reduced, the implementation of measures for monitoring and managing the production, use, and disposal of these pollutants, and the encouragement of global cooperation to help countries adhere to the treaty. The convention aims to curb the release of these toxic substances to mitigate their harmful effects on health and ecosystems.

1.1.1.7 UN-REDD

The United Nations collaborative initiative on Reducing Emissions from Deforestation and forest Degradation (UN-REDD) launched in 2008 at a conference in Switzerland, the UN-REDD Programme is an initiative focused on reducing emissions from deforestation and forest degradation while promoting sustainable forest management and enhancing carbon storage in forests. The key aspects of the programme include developing strategies to help countries adopt effective forest conservation and carbon reduction measures, involving local communities and indigenous groups in forest management, and providing financial and technical support to developing nations. The programme aimed to play a crucial role in global climate change mitigation by addressing deforestation and improving forest health.

1.1.1.8 Nagoya Protocol

The Nagoya Protocol held in 2010 at the Conference of the Parties to the Convention on Biological Diversity (CBD) in Nagoya, Japan, the protocol focused on the fair and equitable sharing of benefits from the use of genetic resources. The major resolutions and highlights of the protocol include creating a legal framework for accessing genetic resources and ensuring that benefits derived from them are fairly shared. It mandates prior informed consent and mutually agreed terms between providers and users of these resources. Additionally, the protocol promotes capacity-building and technology transfer to assist developing countries in implementing its provisions. Overall, it aimed to support global biodiversity conservation and ensure that the benefits of genetic resources are distributed fairly.

1.1.1.9 Minamata Convention

The Minamata Convention was held on 2013 at a conference in Kumamoto, Japan, the convention is an international treaty aimed at safeguarding human health and the environment from the dangers of mercury pollution. The key resolutions and highlights of the convention include phasing out the production and use of mercury-containing products, minimizing mercury emissions from industrial activities and waste management, and regulating artisanal and small-scale gold mining practices. The convention also focused on providing financial and technical assistance to countries affected by mercury pollution to help them manage and reduce its impact. The primary goal was to lower mercury exposure and mitigate its adverse effects on health and the environment.

1.1.1.10 COP21 (Paris Agreement)

The Paris Agreement happened in 2015 during the 21st Conference of the Parties (COP21) at Paris, France. This pivotal accord signifies a major global commitment to addressing climate change. The key outcomes include the target to limit the rise in global temperatures to well below 2 °C above pre-industrial levels, with efforts aimed at restricting the increase to 1.5 °C. The agreement also establishes a system for countries to set, update, and enhance their Nationally Determined Contributions (NDCs), which defined their emission reduction goals. Additionally, it underscores the need for financial and technological support to assist developing nations in adapting to climate change and transitioning to low-carbon economies. The Paris Agreement represents a crucial step in fostering global cooperation and accountability in the fight against climate change.

1.1.1.11 Kigali Amendment

The Kigali Amendment in 2016 at the 28th Meeting of the Parties to the Montreal Protocol in Kigali, Rwanda, focused on combating climate change by phasing out hydro fluorocarbons (HFCs). These substances, commonly used in refrigeration and air conditioning, are significant greenhouse gases. Key resolutions of the amendment include the goal of reducing HFC production and consumption by about 80–85% from baseline levels by 2045, with varying schedules for developed and developing nations. It also highlighted the need for financial and technical support to help countries transition to alternative, less harmful technologies. The Kigali Amendment marks a crucial advance in global climate action by addressing the impact of HFCs on global warming.

1.1.1.12 COP24

The COP24 took place in 2018 at Katowice, Poland, and it was pivotal for advancing the implementation of the Paris Agreement. A major outcome was the finalization of the Paris Agreement Rulebook, which outlines detailed procedures for countries to follow in executing their climate pledges. The key resolutions included the establishment of a comprehensive framework for transparency and accountability in tracking emissions and progress on Nationally Determined Contributions (NDCs), the adoption of rules for international carbon markets, and the strengthening of financial support mechanisms for developing nations. The conference also underscored the need for increased ambition and prompt action to achieve the Paris Agreement's climate goals.

1.1.1.13 COP25

The COP25 took place in 2019 at Madrid, Spain, and it was centred on advancing the global response to climate change and progressing the Paris Agreement's implementation. A key focus was on finalizing the rules for international carbon markets, particularly Article 6, which deals with global carbon trading and collaborative efforts. While the conference did not complete these negotiations, it highlighted the urgent need for stronger commitments to limit global temperature rise to well below 2 °C, with aspirations to reach 1.5 °C. COP25 also reiterated the commitment of developed nations to provide $100 billion annually by 2020 to support climate action in developing countries and stressed the importance of bolstering adaptation and mitigation measures.

1.1.1.14 COP26

The COP26 took place in 2021 at Glasgow, United Kingdom, and it was a pivotal event for advancing global climate efforts and enforcing the Paris Agreement. The conference underscored the critical need for urgent climate action and equitable solutions. The major outcomes included a push for countries to enhance their Nationally Determined Contributions (NDCs) and elevate their climate commitments, with a focus on restricting global temperature rise to 1.5 °C. The key achievements also included finalizing the rules for international carbon markets under Article 6 of the Paris Agreement, emphasizing a just transition for those affected by climate policies. Additionally, COP26 addressed financing for climate action and adaptation in developing countries and included agreements to reduce coal use and methane emissions.

1.1.1.15 COP27

The COP27 was held in 2022 at Sharm El Sheikh, Egypt, and played a key role in advancing global climate initiatives and addressing climate impacts. A major outcome was the ground breaking decision to create and implement a Loss and Damage Fund to provide financial assistance to countries experiencing severe climate-related impacts. The conference also stressed the importance of boosting climate adaptation and resilience, particularly for vulnerable developing nations. In addition, COP27 focused on increasing climate finance and reinforcing commitments to cut greenhouse gas emissions, supporting the objectives of the Paris Agreement.

1.1.1.16 COP28

The COP28 took place in 2023 at Dubai, United Arab Emirates, and was crucial for intensifying global climate efforts and tackling climate change impacts. The conference underscored the immediate need for enhanced climate action and adaptation strategies. A key decision was to increase financial support for nations significantly affected by climate change, addressing both short-term and long-term challenges. Additionally, COP28 stressed the importance of accelerating the shift to renewable energy and reducing emissions across different sectors. The event also called for strengthened international collaboration and greater ambition in national climate commitments to meet the Paris Agreement's objectives.

1.1.2 India in Demand-Side Management of Emission

India, ranked as the world's third-largest emitter of greenhouse gases, has undertaken significant efforts to combat climate change and reduce its carbon footprint. India's climate strategy predominantly emphasizes expanding renewable energy sources, there is comparatively less emphasis in available literature on managing demand-side aspects and consumer behaviour. Nevertheless, India's climate diplomacy capitalizes on its soft power to foster global confidence and steward common resources (Teske 2019), establishing its role as a growing influence in global climate governance. This multifaceted approach reflects India's endeavour to harmonize immediate domestic needs with the long-term benefits of combating climate change, taking into account both domestic constituents and international commitments (Zhang et al. 2023). India's approach to climate change and reducing carbon emissions has undergone significant transformation over time, moving from a defensive stance to a more proactive role both at home and on the global stage. Initially, India focused on forging alliances with other developing nations to secure commitments from developed countries on emission reductions, financial support, and technology transfers (Sengupta 2019). However, as India's economic stature grew, its climate policy evolved. The country began to demonstrate greater flexibility in international negotiations, offering voluntary commitments while still advocating for differentiated responsibilities between developed and developing countries (Sengupta 2019; Betz 2014). At the domestic level, India has increasingly integrated climate considerations into its development strategies, motivated by concerns such as energy security, susceptibility to climate-related events, and the potential for fostering a competitive green technology sector (Saryal 2018; Dubash et al. 2018). In recent years, India has emerged as a significant leader in climate action, aligning its international pledges with domestic policies (Chen et al. 2023).

Internationally, India has consistently emphasized holding developed countries primarily accountable while positioning itself as a developing nation. However, domestically, India has taken on increased responsibility by implementing climate policies aligned with its national interests. This dual strategy enables India to independently shape its role in climate governance, balancing developmental imperatives with environmental considerations (Prys-Hansen 2022). India's climate and energy policies are shaped by both internal priorities and external pressures. This nuanced approach reflects India's endeavour to navigate the complexities of economic growth, energy security, and environmental sustainability while asserting its growing influence on the global stage (Prys-Hansen 2022).

1.1.3 Climate Change Action Plans: Indian Perspective

India has been actively pursuing climate policies and actions aimed at reducing greenhouse gas emissions, particularly at the individual and household levels. The country understands the importance of involving citizens in climate action, as individual behaviours have a significant impact on overall emissions. On the demand side, India has implemented various regulations to mitigate climate change. The Energy Conservation Building Code (ECBC) sets energy efficiency standards for commercial buildings, and the Bureau of Energy Efficiency (BEE) establishes standards for appliances and equipment to reduce energy consumption. Additionally, the government introduced the National Electric Mobility Mission Plan (NEMMP) to promote electric vehicles (EVs), aiming to decrease reliance on fossil fuels and reduce emissions from the transportation sector (NITI Aayog 2019). India faces significant challenges due to climate change, including increased extreme weather events, retreating Himalayan glaciers, and rising sea levels. In response, India has implemented various measures focusing on both mitigation and adaptation. The key initiative includes the National Action Plan on Climate Change (NAPCC), which addresses solar energy, energy efficiency, sustainable agriculture, and water conservation, among other areas. India also co-launched the International Solar Alliance (ISA) to promote solar energy globally and aims to achieve 100 GW of solar power capacity by 2022 through the National Solar Mission. Additionally, the Green India Mission seeks to enhance carbon sinks in forests, and climate-resilient agriculture is being promoted to strengthen the farming sector's resilience. Internationally, India has made significant commitments, including reducing the emissions intensity of GDP by 33–35% by 2030 under the Paris Agreement and achieving net-zero emissions by 2070, as announced at COP26. The future goals include advancing the National Hydrogen Mission, enhancing climate resilience, and promoting electric mobility. Through these efforts, India aims to balance development with environmental sustainability, playing a key role in the global fight against climate change.

The evolution of global climate negotiations from COP24 to COP28 (2018–2023) highlights a shift towards more tangible actions and commitments in tackling climate change. This period saw major advancements in policy frameworks, financial mechanisms, and national commitments, with India playing a pivotal role in shaping the discussions. The Katowice Climate Package, adopted at COP24, established a comprehensive rulebook for implementing the Paris Agreement, laying a foundation for future climate actions (UNFCCC 2018). This was a crucial step in operationalizing the Paris Agreement, providing clear guidelines for countries to follow in their climate mitigation and adaptation efforts. Subsequent COPs built on this foundation, addressing key issues such as carbon markets, loss and damage, and enhanced national commitments. A recurring theme throughout these conferences was the focus on climate finance. The repeated calls for developed countries to meet the $100 billion annual commitment highlighted the challenges in mobilizing sufficient financial resources for climate action in developing countries (Bhowmick 2021). This financial shortfall has been a significant barrier to implementing ambitious climate policies in many developing nations, including India.

India's participation in these COPs was marked by a consistent advocacy for climate justice and equity. At COP24, India emphasized its ambitious renewable energy targets and afforestation efforts, showcasing its dedication to climate action while addressing its development needs (Ministry of Environment, Forest and Climate Change 2018). This approach aligns with the principle of "common but differentiated responsibilities" outlined in the United Nations Framework Convention on Climate Change (UNFCCC). A significant milestone in India's climate commitments came at COP26 with the announcement of the "Panchamrit" plan. This five-point agenda included ambitious goals such as achieving net-zero emissions by 2070 and increasing non-fossil energy capacity to 500 GW by 2030 (Dubash et al. 2022). These commitments represented a major shift in India's climate policy, aligning more closely with global efforts to limit temperature rise to 1.5 °C above pre-industrial levels.

The focus on adaptation gained prominence in the later COPs, especially at COP27 and COP28. The agreement to establish a loss and damage fund at COP27 was a significant development, recognizing the disproportionate impact of climate change on vulnerable countries (UNFCCC 2022). India's emphasis on climate-resilient infrastructure and sustainable agriculture at these conferences underscores its understanding of the importance of adaptation in its climate strategy (Jha 2023). The first global stocktake conducted at COP28 provided a critical assessment of collective progress towards the Paris Agreement goals, underscoring the urgency of enhancing national commitments and accelerating the energy transition (UNFCCC 2023). India's submission of updated NDCs at this conference illustrated its dedication to increasing its climate ambition in line with global expectations. Throughout these COPs, India maintained a delicate balance between its climate commitments and development needs. The country consistently emphasized the necessity for enhanced support from developed nations in terms of finance and technology transfer (Ghosh 2023). This stance reflects the challenges faced by developing countries in implementing ambitious climate policies while addressing pressing socioeconomic needs.

1.2 Conclusion

Global and national climate change conventions are essential in tackling the complex challenge of climate change by fostering international cooperation, setting binding commitments, and creating frameworks for sustainable practices. Active participation in these conventions enables countries to share expertise, mobilize resources, and adopt innovative technologies for climate mitigation and adaptation. National initiatives, like those in India, complement global efforts by addressing local needs through renewable energy adoption, sustainable agriculture, and enhanced climate resilience. India's commitments under the Paris Agreement and COP26, along with its goal of net-zero emissions by 2070, illustrate how countries can align domestic priorities with global climate objectives. The collaboration between global conventions and national actions is key to effectively addressing climate change and ensuring a resilient and sustainable future.

Household Carbon Footprint (HCF) Assessment in India: Approaches and Challenges

2

The carbon, land, and water footprint can provide important insights in understanding of anthropogenic impact on the ecosystem. Household consumption patterns have been identified as one of the important factors in influencing greenhouse gas emissions and considered to be critical for developing climate policy. IPCC (2007) suggests change in consumption patterns as a possible approach to alleviate the effects of climate change. Various socioeconomic, lifestyle and demographic characteristics such as household size, household income, food habits, mobility choices, waste management, home appliances, gender, education, and employment significantly influence household carbon footprint. Carbon footprint is calculated by adding the emissions of different greenhouse gases (GHGs) resulting from lifecycle (cradle to grave) of a product or service (material sourcing, manufacturing, use phase, and end-of-life disposal). Unilever in one of its studies had estimated that approximately 70% of GHGs footprint depends on which product consumers choose and whether they use or dispose the products in a sustainable way.

While direct emissions from production of energy, other products and services are an integral part of the IPCC mitigation discourse, less emphasis has been given to indirect emissions by advocating lifestyle and end-use consumption changes. A number of studies on measurement of household carbon footprint in developed countries are available but very limited studies have been conducted in developing countries to determine the carbon footprint of households. The studies have employed various perspectives and methods to examine resource use and wastes associated with consumption studies. Many studies also have linked household expenditure surveys with environmental intensities to show the distribution of environmental footprints. The member countries of the Organisation for Economic Co-operation and Development (OECD) have designed and implemented policy frameworks to reduce environmental impacts from household activities. The policies primarily aim to transform household decision-making processes to adopt and promote sustainable behaviours like energy conservation methods or waste

S. Nautiyal et al., *Determinants and Stratification of Household Carbon Footprint*, https://doi.org/10.1007/978-981-96-2739-4_2

recycling. Other measures like imposing standards, taxes, penalty, etc., are also instituted to influence consumers' choices in the market.

According to the World Resources Institute (WRI) report (2014) India is the world's third largest contributor to anthropogenic greenhouse gas (GHG) emissions although per capita emission is comparatively lower. India has been in the midst of demographic, economic, and technological transitions. Factors like growing aspirational middle class with better disposable incomes, rapid rate of urbanization, and better access to resources have been contributing towards increasing emissions levels, both per capita and per household over time. India is expected to become one of the fastest growing consumer appliances markets driven primarily by rapid urbanization, household electrification, growing per capita income, and female participation in the workforce (PWC 2018).

CO_2 emissions from energy consumption have increased significantly in emerging economies. Residential sector in India accounted for 24% of the total electricity consumption in 2017–18 (Energy Statistics Report 2020). India's annual per capita energy consumption is 0.6 tonnes of oil equivalent, about a third of the world average according to the Economic Survey of 2018–19 (MoF 2019). Around 750 million people in India gained access to electricity between 2000 and 2019. In March 2019, India had achieved full electrification of almost all households (IEA 2020). Projection done by NITI Aayog indicates that the electricity consumption for the residential sector is expected to increase 6–13 times by 2047.

The transport sector is responsible for 8 Gt per year of CO_2 global emissions and road transport alone is responsible for about 74% of the sector's total emissions (IEA 2018). In India, the transport sector is the third largest energy consumer at 17% of Total Final Consumption (TFC) and 26% of total emission in 2017. Considerable growth from 0.3 Gt CO_2 in 2017 to over 0.7 Gt CO_2 in 2040 is expected primarily from road transport sector in India. There are approximately 30 Light Duty Vehicle (LDVs) per 1000 individuals (IEA 2019).

Food accounts for a major share of a household's carbon footprint. Household emissions from food are determined by the volume and composition of food consumed and the environmental intensity of that food. United Nations estimates that the annual carbon footprint of global food waste is 4.4 gigatons of CO_2 equivalent. The majority of the GHG emissions in Indian diets are accounted for by rice- and animal-source foods. Firewood is still used by over 500 million people across India and is sometimes the only domestically available and affordable source of energy, especially for low-income households. The IPCC Special Report on 1.5 °C (IPCC SR1.5) states that lack of access to clean and affordable energy for cooking is a major policy concern in many countries. Fuel burning is estimated to add the equivalent of another 8–16% of the global warming caused by CO_2 (Bailis et al. 2015).

The United Nations estimates that the share of the world's population that lives in cities will double from 30% in 1950 to 60% in 2030 (UN 2014). In addition, urbanization, especially in the developing world, has occurred at an unprecedented pace over the past few decades. The impact of urbanization on energy use of fossil fuels has increased the carbon levels in the atmosphere as urban residents usually consume more goods and services than rural residents. The urbanization–emission nexus in India needs special attention as the county has already witnessed considerable urbanization and expected to continue with regard to smart cities initiative of the government. Nearly 70% of India's population lives in rural areas; this number is unlikely to change in coming decades. Hence, rural India has the potential to contribute in carbon emission reduction provided a strategic and doable framework is in place for low carbon development and mitigation of climate change, given a global climate change dilemma in view.

Promoting sustainable lifestyles can be an essential strategy towards achieving the 12th UN's Sustainable Development Goal (SDG) of Responsible Consumption and Production. The SDG 12 is a reflection and acknowledgement of the fact that resource-intensive lifestyles are detrimental for the achieving the concept of living well within earth's limits. Household activities, including consumption behaviours, are interconnected by social-economic factors, a holistic evaluation of the household practices is essential to determine influencing factors, alternate lifestyle options, and rebound effects. Further the research findings are expected to contribute towards actions for overall development of low-income households and thereby aid in achieving SDG-1, 2, and 3. The recommendations from the study will also suggest pathways for climate action from individual household level to state and national level (SDG-13).

2.1 Past Estimates of Carbon Footprint in India

Calculators for estimating individual and household carbon footprint are flooded in different platforms, specifically in web-portals of governmental, and non-governmental environmental organizations. The results of those of those calculators vary to different extents although the emission factors used are globally accepted because of different methodological approaches, inclusion, and exclusion of items. There are several scientific research studies in this regard to develop guidelines for accurate assessment of GHG emissions. In Indian context, to address the diversity in geophysical and socioeconomic variables controlling consumption pattern, several attempts have been made by scholars to estimate carbon footprints of individuals, households, and cities; these include several approaches, viz. average balanced diet: fuel consumption, transportation, raw material production, processing (Pathak et al. 2010); domestic carbon footprint: energy used/fuel (Ramachandra et al. 2014); environmental effects of dietary change: national availability indicators, planetary boundary framework metrics (Gill et al. 2018); household size and level of urbanization changing consumption behaviour as factors affecting carbon emission (Sikdar and Mukhopadhyay 2016); energy use and carbon emission along rural-urban gradient and urban hierarchy (Sethi 2015); CFP using consumption of goods and services. Some of these studies have addressed either any one or a few items contributing to household carbon footprint. Consumer expenditures and their corresponding carbon-intensity was the key data used for estimation of total household carbon footprint in previous studies. Among the most recent studies in India regarding estimation of carbon footprint, both are very elaborative and data extensive research based on secondary data from NSSO and IMS. Within this study, Lee et al. (2021) calculated and compared the household carbon footprints of 623 districts in India, based on micro-consumption data from 203,313 households and explain their variation by economic, cultural, and demographic factors. The results show eradication of extreme poverty does not conflict with ambitious climate change mitigation in India.

2.2 About the Study

The present study was assigned to Institute for Social and Economic Change (ISEC, Bengaluru, India) as a collaborative project with PIKS and a part of a broader project "Supporting the Institutionalisation of Capacities on Climate Change Studies and Actions," implemented by MoEFCC and GIZ. It is expected to contribute towards network building, knowledge exchange, and strengthening capacities for enhanced science-policy-practice. The study has been designed for *conducting primary survey on consumers' attitudes towards low carbon lifestyle*. The assignment includes the following objectives:

(a) Quantification of carbon footprint of Indian households.
(b) Identify the increase in consumption of energy/emission intensive goods/services with an increase in household income.
(c) Identify pathways for sustainable lifestyles including those in traditional lifestyles.
(d) Assess the impact of COVID-19 led lockdown on consumption behaviour of different types of households in India.

The objectives also include ground-truthing of national estimates and providing concrete suggestions to policy makers (MoEFCC) to mainstream mitigation actions identified on the basis of sample survey in target sectors. The proposal intended collection of primary data from 1500 households from two geographical region of India through random sampling considering inclusion of variability pertaining to economic and social well-being. This comprehensive top-down and bottom-up analysis will identify the sectoral increase in HCF and identify potential pathways for HCF reduction. This will aid the MoEFCC to take informed science-based actions in these sectors. The authentic empirical data set is also expected to provide a direction to adopt required changes in the frameworks of national level data (e.g. NSSO) collection. Further, based on the discussion in the inception meeting and looking at the aggravating situation of COVID-19 pandemic, restructuring of methodology took place to attenuate the extensive travel and fieldwork and to achieve the broader goal of the project. Thirteen researchers from network organizations were trained from network organizations to collect data locally from 13 locations (Fig. 2.2).

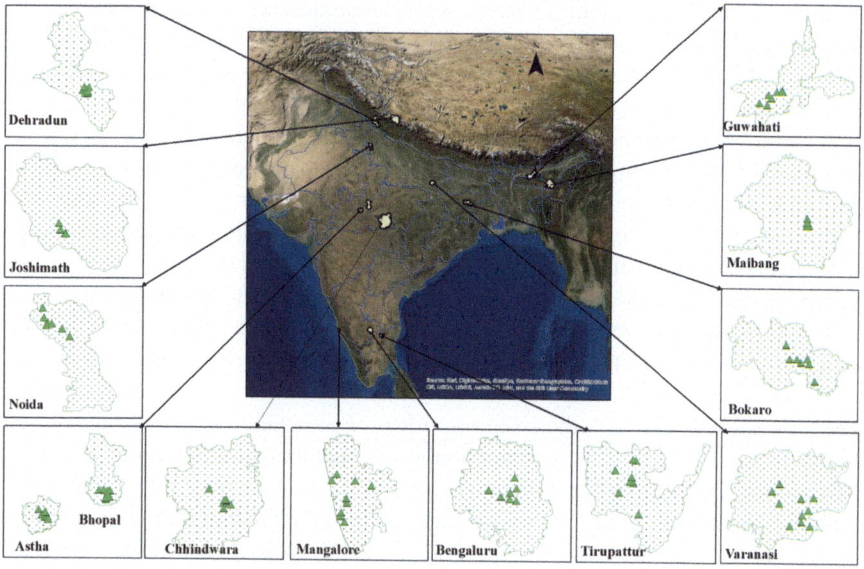

Fig. 2.1 Map showing selected locations

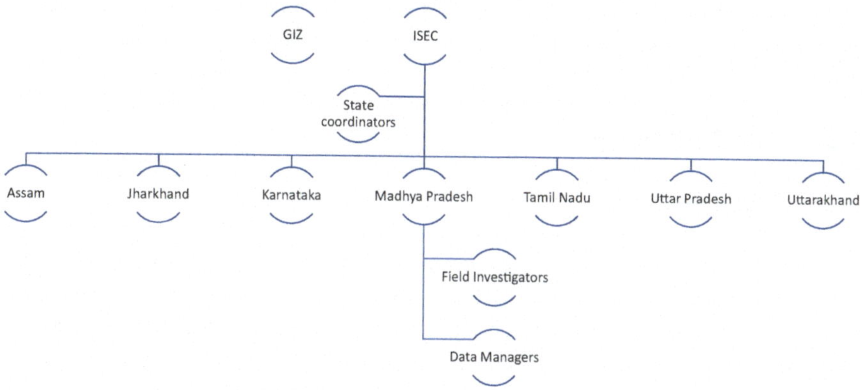

Fig. 2.2 Organizational setup

While the economic liberalization initiated in 1991 and programmes addressing energy poverty in India have lifted many Indian households from absolute poverty and have created millions of middle-class families (mostly centred in cities) who demand more energy for their sustenance. In fact, the growing middle class—with up to 300 million people—implies an acute shift in the consumption amount and type of goods and services which is further triggering the increase of the embodied emissions. The contribution of household carbon is significant which is primarily guided by income and consumer behaviour. Variations in household emission in rural and urban India among different socioeconomic classes are also noticeable. Studies also suggest that high emission doesn't necessarily bring better well-being in households. Unlike developed countries, developing countries have a reverse pattern of emission with the degree of urbanization where households in urban areas emit more. A comprehensive understanding of emission pattern across various groups and sectoral emissions is required to identify pathways for emission reduction which can contribute towards mitigation planning through both policy and awareness building. Another important aspect in implementing mitigation strategy is acceptance of mitigation pathways by different social and economic sector, considering households' development goals and varied social and cultural norms.

While mitigation can be addressed from a top-down perspective through increased investments on a large scale for renewables, the role of households as end-users and demand-side mitigation often remained unaddressed in Indian context. The joint study focuses on the issue of mitigation at a national level solely from a demand-side perspective. It encompasses all the sectors that are covered in India's NDCs since household's consumption behaviour dictates the need for a given good/service in the market. This, on the one hand, impacts the GDP of the country and on the respective sectoral emissions on the other hand. For instance, a given household cannot have a direct influence on the type of electricity (i.e., coal powered or renewable) it receives but can decrease its HCF by minimizing electricity consumption through increased awareness and/or through installing energy-efficient appliances. The planned primary survey addresses this aspect of Indian households and identifies pathways for sustainable lifestyles.

2.3 Methodology

2.3.1 Approach

The study necessitates evaluation of household carbon emissions to understand the extent of nation-wide emissions. The methodological design adopted is cross-sectional with consultative approach. The primary objective of the study is to determine the carbon emissions of households and their changing lifestyle whereas emissions comparison among states and regions (Rural/Urban/Peri-Urban) based on distinctive variables are ancillary to the primary objective. For the given objectives, it was required to have a descriptive as well as perspective response to frame the study. Several ingenious variables were created within the questionnaire to seek information on carbon emissions from the people's standpoint in addition to their consumption and expenditure pattern which has an impact upon the emissions, individually and as a household. Eventually, an analysis was performed by comparing the values obtained with the national estimates, at length.

2.3.2 Research Design

2.3.2.1 Sampling

A single-stage cluster sampling technique was taken into consideration for the household study. The target population was the individual households across three zones in every selected study area. All the members of the selected households (Cluster) were also included as household has been taken as the respondent unit of the survey. The selection of the sample is prioritized upon the different agro-climatic zones and degree of urbanization (Table 2.1). The sample was drawn from three zones of each location, viz. urban, peri-urban, and rural. The zones are defined on the basis of distance from the city core, administrative boundary, population density, and accessibility.

2.3.2.2 Study Area

See Table 2.1.

In India, agricultural landscapes are classified into agro-ecological zones, which are curved out of climate zones, and further 20 regions are identified based on physiographic features, soil characteristics, bio-climatic features and length of growing periods. National Bureau of Soil Survey and Land Use planning (ICAR) have

Table 2.1 Description of the locations

S. no	Location	States	Agro-climatic zone	Location of sample coverage Rural (R), Peri-urban (PU), Urban (U)
1.	Tirupattur	Tamil Nadu	Southern Plateau and Hills	R, PU, U
2.	Maibang	Assam	Eastern Himalayas	R, PU, U
3.	Guwahati	Assam	Eastern Himalayas	R, PU, U
4.	Varanasi	Uttar Pradesh	Middle Gangetic Plain	R, PU, U
5.	Noida	NCR	Upper/Trans Gangetic Plain	PU, U
6.	Bengaluru	Karnataka	Southern Plateau and Hills	U
7.	Mangalore	Karnataka	West Coast Plains and Hills	R, PU, U
8.	Bokaro	Jharkhand	Eastern Plateau and Hills	R, PU, U
9.	Bhopal	Madhya Pradesh	Central Plateau and Hills	R, PU, U
10.	Chhindwara	Madhya Pradesh	Central Plateau and Hills	R, PU, U
11.	Ashta	Madhya Pradesh	Western Plateau and Hills	R
12.	Joshimath	Uttarakhand	Western Himalayas	R
13.	Dehradun	Uttarakhand	Western Himalayas	R, PU

sub-divided the regions into 60 agro-ecological sub-regions in India. The physiographic and climatic heterogeneity, cultural diversity, and agricultural practices have contributed towards a diverse agro-biodiversity country. These highly diverse ecosystems provide conditions for the adaptation of a large number of both indigenous and exotic crop plants bringing a revolution in the growth of agriculture and agro-biodiversity (Shetty et al. 2020). India is predominantly an agricultural country and consumptions are mostly determined by local produces. Landscape change, change in consumer behaviour, climate change, market-oriented cultivation, etc. are some of the determinants of agro-ecological characteristics and vice-versa. Therefore, selection of study areas has been done to cover agro-ecological diversity of the country. It is expected to help in developing place-specific knowledge of emission reduction pathways and decision-making. The survey was carried out in 13 different locations (Dehradun, Joshimath, Noida, Bengaluru, Mangalore, Tirupattur, Varanasi, Bokaro, Guwahati, Maibang, Ashta, Chhindwara, and Bhopal) from nine agro-climatic zones of India (Fig. 2.1).

2.3.2.3 Detailed Description of the Study Area

Each area was scientifically selected based on their geoclimatic locations, agricultural, ecological, social, and economic settings to develop an inclusive sample representative of diversity of India. The weightage of 150 households from rural/peri-urban/urban was evenly distributed for locations: Tirupattur, Maibang, Guwahati, Varanasi, Mangalore, Bokaro Steel City, Bhopal, Chhindwara whereas Ashta, Joshimath, and Dehradun are exclusively rural locations, Noida encompasses only urban and peri-urban setting and Bengaluru is exclusively urban location (Fig. 2.3). The total sample size comprises of 1950 households. The habitants of different agro-climatic zones will have different climatic settings, lifestyle, energy demand, consumption, and emission pattern.

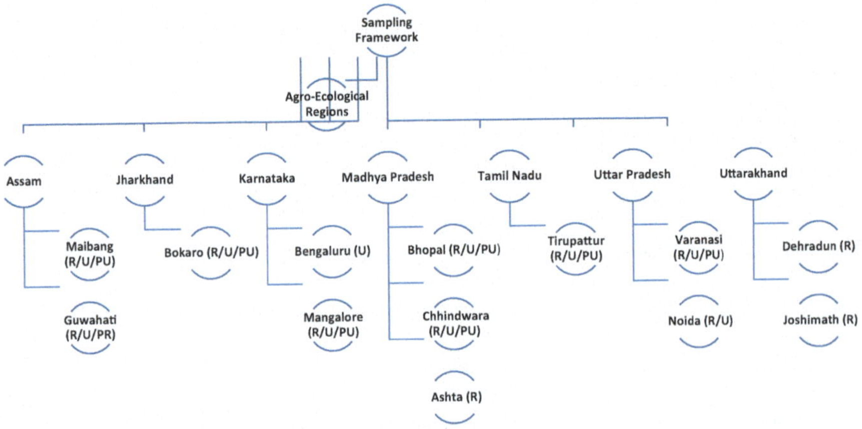

Fig. 2.3 Structural arrangement of data collection in study region

Tirupattur

Tirupattur is a town and it is the head quarter of Tirupattur district of Tamil Nadu state. This town is also called as sandalwood town due to plentiful availability of sandalwoods in surrounding hills of Tirupattur town. This city lies between 12.50°N latitude and 78.60°E longitude with an average elevation of 387 m ASL. Tirupattur town shows tropical savannah climate with moderate weather conditions throughout the year. Annual average maximum and minimum temperatures of Tirupattur are 37.0 °C to 16.1 °C, respectively. Average annual rainfall of this town is 877 mm. As per census 2011, Tirupattur population was 19,487 and the literacy rate of this town was 76.22%. Tirupattur falls under Southern Plateau and Hills agro-climatic zone.

Maibang

Maibang is the part of Dima Hasao district in the state of Assam, India. This small town lies between 25°17′43″N longitude and 93°08′22″E latitude with an average elevation of 355 m. It is located in a hilly area at the bank of Mahur River with an area of 10 km^2. This town is home for diversified tribal community groups that contribute to social fabric of the district. Thirteen tribal groups in Dima Hasao district, namely Dimasa, Kuki, Zeme, Hmar, Karbis, Khelma, Hrangkhol, and Biates. As per the census 2011, population of the Maibang is 20,664 and population density of 2100/km^2. Tourism and transportation are the major sources of the economy of the Maibang town. Maibang town is covered within Eastern Himalayan agro-climatic zone.

Guwahati

Guwahati is one of the metropolitan cities, and it is located in the north eastern region of the India. This city is located on the Brahmaputra River bank and lies between latitude 26°10′20″N and longitude 91°44′45″E. Guwahati is one of the fastest growing cities of India. As per 2011 census, population of Guwahati was 9,57,352 and the average literacy rate was 91.47%. This city located towards the south-eastern side of Kamrup district surrounded by Darrang and Marigaon districts in the East, Goalpara and Barpeta districts in the West, Nalbari district in the North, and Meghalaya State in the south. This city is situated on the undulating topography with the varying elevation from 49.5 to 55.5 m above mean sea level. Also, this city has small hillocks, namely Chunsali hill (293 m), Nabagrah hill (217 m), Nilachal hill (193 m), and Sarania hill (193 m). Guwahati has a humid subtropical climate with an average annual temperature of 23 °C. The overall habitat type in the study area mainly comprises of forest patches, wetlands, agricultural lands grasslands, scrublands, plantations, human settlements, and commercial areas. The forest patches are of a moist deciduous type. Sal formations mixed with patches of evergreen and bamboo formations can be observed in southern periphery. The manufacturing sector plays a crucial role in economy of Guwahati and it contributes ample share to the city's economy. Guwahati city is covered within Eastern Himalayan agro-climatic zone.

Varanasi

Varanasi is situated in eastern part of Uttar Pradesh with 25° 14–25° 23.5 N latitude and 82° 56–83° 03 E longitude with an average elevation of 80.71 m ASL in Middle Gangetic Plain. With urban agglomeration, this city covers an area of 112.26 km^2. This city is considered as the oldest city of the world, and it is located along the left crescent-shaped bank of the Ganga River. As per 2011 census, Urban Agglomeration of Varanasi had a population of 14, 35,113 and the literacy rate is 77%. This town experiences a humid subtropical climate with large variations between seasons. In summer, temperature ranges from 22.5 to 46 °C and in winter particularly temperatures fall down below 5 °C due to cold waves from the Himalayan region. The annual average rainfall is 1110 mm. Manufacturing, tourism, trade and commerce, transport and communication, and agriculture are the major economic sources of the Varanasi town. Particularly in manufacturing industries, spinning and weaving are the major activities.

Noida

Greater Noida located in Gautam Buddha Nagar district of Uttar Pradesh with an area of 380 km^2 lies between latitude 28.4744°N and longitude 77.5040°E. As per census 2011, population of Greater Noida is 107, 676 and the literacy rate was 86.54%. This city has tropical savannah climate. Weather is extreme hot and dry during summer, hot and humid during monsoons, extreme cold during winter, and pleasant weather can be observed during spring and autumn. The setup of several corporate companies and industries are the major income sources of the Greater Noida. Noida is part of Upper/trans Gangetic Plain.

Bengaluru

Bengaluru is the capital and largest city of Karnataka state. It is located in the south-east of Karnataka state between latitude 13°24′24.86″N and 13°3′39.414″N and between longitude 77°33′42.701″E and 77°36′28.573″E. The average elevation of Bengaluru is 900 m MSL. An urban and rural area of Bengaluru covering an area is about 2,191 km^2. Population growth of Bengaluru is very rapid and the city has grown spatially more than 10 times since independence. As per 2011 census, the population of Bengaluru city is more than 8 million which made Bengaluru as the third most populous city in India. Topography of Bengaluru varies from area to area. Uneven landscapes with hills and valleys are observed in southern part, level plateau is observed in northern part, plain minor undulations are observed in eastern portions of Bengaluru and rugged topography can be observed in west portions of Bengaluru. Bengaluru city shows tropical savannah climate with both dry and wet seasons. The average low temperature of this city is 15.1 °C and the highest average temperature is 35 °C. The mean annual rainfall of Bengaluru city is 859.6 mm. Economy of Bengaluru plays a crucial role in Karnataka's economic growth; also this city contributes a major portion of economy particularly from IT exports. With more 10% economic growth, Bengaluru stands in the second position among metropolitan cities of India. The agro-climatic zone that covers Bengaluru is Southern Plateau and Hills.

Mangalore

Mangalore (Mangaluru) is the administrative nerve centre and largest city in Dakshina Kannada district of Karnataka state, in southern India. Mangalore city is located between 12.9141°N latitude and 74.8560°E longitude with an area of 132.4 km^2. This city is highly urbanized and one of the cosmopolitan non-metro cities in India. It has a modern port which is located 10 km north of Mangalore; this port is India's ninth largest cargo handling port. Mangalore city is one of the highly urbanized cities of India with more than 4 million urban populations as per 2011 census (Census 2011). This city is located between the Arabian Sea and the Western Ghats mountain ranges, and it falls under coastal agro-climatic zone. Mangalore city urban agglomeration extends from Ullal in the south to Mulki in the north, covering a distance of over 40 km. Mangalore city's landscape is characterized by rolling hills, coconut palms, hard red-clay tiled-roof buildings, and freshwater streams. The climate of Mangalore is a tropical monsoon and the main rainy season starts from June and it will continue till September. However, heavy rainfall occurs in the month of July. Annual average rainfall of Mangalore is 3672 mm (pre monsoon 251 mm; southwest monsoon 3103 mm; northeast monsoon 318 mm rainfall). High humidity can be observed during June, July, and August months. The major economic activity in Mangalore is commercial, industrial, port-related activities, and agriculture processing.

Bokaro

Bokaro is also called as steel city; it is the administrative headquarters of Bokaro district of Jharkhand state. This city lies between 23°40′N and 23°67′N latitude and between 86°09′E and 86° 15′E longitude with an average elevation of 210 m ASL and has an area of 183 km^2. As per 2011 census, the population of Bokaro is 5, 63,413 with 84.87% literacy rate. This city lies in the tropical area and this city experiences five seasons, viz. summer, rainy, autumn, winter, and spring seasons. Summer season starts from April and it will continue till mid-June. Maximum and minimum temperatures of summer season are 30–40 °C. Similarly, winter temperature ranges from 8 °C to 20 °C. The annual average rainfall of Bokaro city varies from 140 to 180 cm. The main economic source of this city is integrated steel plant established by Steel Authority of India.

Bhopal

Bhopal city is the capital city of Madhya Pradesh state of India. It is located in the southern part of Bhopal District of Madhya Pradesh and lies between latitude 23° 05 N and 23° 54 'N and longitude 77° 10'E and 77° 40'E with an average elevation of 500 m MSL. Bhopal shows a humid subtropical climate, with dry, cool winters, humid monsoon, and hot summer season. Annual average temperature of Bhopal is 30 °C and during summer season particularly in the month of May it exceeds more than 40 °C. According to the 2011 census, the population of the Bhopal city is 18, 86,100. The literacy rate was 85.24%. Economic activity of Bhopal includes different industries (cotton, chemical, electrical goods, etc.,), large retail business, and handicrafts. Bharat Heavy Electricals Limited (BHEL) which is one of the largest engineering and manufacturing enterprises in India unit occupies a large area of eastern part of the city and it provides a livelihood to majority of peoples of Bhopal. Also, Bhopal was selected to be developed as a smart city in first phase of Smart Cities Mission. Also, in three continuous consecutive years Bhopal was rated as the cleanest state capital in India.

Chhindwara

Chhindwara is one of the largest cities in the Satpura Range, and this city is an administrative headquarters of Chhindwara district of Madhya Pradesh state and also this city is a Municipal Corporation of Chhindwara district. This city lies between 22.07°N latitude and 78.93°E longitude with an average elevation of 675 m ASL. As per 2011 census, population of urban agglomeration of this city was 1, 90,008 and literacy rate of this city is 89.25%. This city experiences large variations in temperature pattern in different seasons. Annual average maximum and minimum temperatures are 30.8 °C and 17.1 °C, respectively. In summer season, the temperature goes up to 47.6 °C (May month) and in winter season temperature falls down to <4 °C. Chhindwara is rich in manganese, mica, limestone silicate rocks, and coal mines. Industries (coal mines) are the main economic income source of the city.

Ashta

Ashta is a small city and municipality of Sehore district of Madhya Pradesh state. This city is lies between 23.0180°N longitude and 76.7160°E latitude with an average elevation of 519 m. As per census 2011, the population of Ashta municipality was 53,184 and the literacy rate of Ashta city was 79.41%. The average annual temperature of Ashta is 25.5 °C and the annual rainfall here is 1126 mm. During summer season in Ashta, maximum temperature reaches up to 40 °C in May month and during winter season temperature will fall down to <12 °C in the month of January.

Dehradun

Dehradun is situated in north-west corner of Uttarakhand state and located between 29° 58′N and 31° 02′ 30″N latitude and between 77° 34′ 45″E and 78° 18′ 30″E longitude with an average altitude of 640 m MSL. The Dehradun district has various types of physical geography from Himalayan Mountains to Plains. As per 2011 census, population of Dehradun is 5,78,420 and the literacy rate of the city is 89.32%. Dehradun shows humid subtropical climate. The average maximum temperature in summer is 36 °C and minimum is 16° however in winter it varies from 2.3 °C and 5 °C. The average annual rainfall of Dehradun is 2183.5 mm. About 87% of the rainfall can be observed during monsoon season (June–September) particularly, substantial rainfall observed in the month of July and August. Tourism is the main source of economy in Dehradun. Also, majority of the population of Dehradun are depended on agriculture.

2.3.2.4 Data Sources

The primary and secondary data sources have been used throughout the study. The primary data was collected through data instruments with predetermined variables to capture the consumption needs of households. As the data for comparison for basic level analysis is either unavailable or obsolete; therefore, it enables us to conduct per capita, zone, and state-wise estimation analysis, and eventually use national estimates for total valuation. The secondary sources include per capita consumption of energy, food, and other resources for consideration but majorly the conversion factors were used to estimate emissions from household consumption of energy, food, transport, and waste among other variables. The secondary sources are enlisted below (Table 2.2):

Data Collection

The combination of both qualitative and quantitative methods were used, i.e., convergent parallel mix method, to obtain data.

Questionnaire used consists of series of questions. The process of gathering data was accomplished by door-to-door and through telephonic conversation. The sections in the questionnaire were organized into the following heads: Family profile, household annual income, household expenditure, land and living space, agriculture and livestock, energy use, household assets, transport, food consumption, and waste generation.

Structured Interview schedule is one of the tools which consists of predetermined set of questions. Section 18 of the schedule was used to elicit responses pertaining to behavioural characteristics and situational dilemma. The set of questions maintains "Willingness to Change" aspect with reference to expenditure on eco-friendly services, subsidies, incentives, and policy suggestions by respondents.

Definition of terms used in the survey:

- The survey unit in this survey is a household. We define household as all the persons living in the same shelter and sharing the same meals.
- The definition of household carbon footprint is a count of the carbon dioxide and other greenhouse gases (the CO_2 equivalent) ascribed to a household due to their activities and the energy used for the goods and services they use.
- The estimations of greenhouse gases will be done from the estimations of consumption of goods, use of services, and practice of any norms or actions by each member of the household.
- All the data and information in the survey should be an aggregate or average (as indicated in the questions) for the entire household.

2.3.2.5 Sampling Framework

In Table 2.3, the total number of households is 1796. Additional 50 households from urban Bengaluru belonging to upper middle- and high-income groups were taken which make the total sample size 1950. The table shows relationship between the socioeconomic variables and relationship between the zones (R, U, PU). Each socioeconomic variable shows percentage distribution of the over the three zones. The major part of the sample comprises of rural areas. The urban zone holds the

Table 2.2 Description of the secondary data sources

Parameters	Source	Year	Remarks
Education: Households by number of literates	www.mhrd.gov.in Census 2011	2011	The data is old to consider when compared to current situation
Income: National State Domestic Product Per capita income	www.mospi.gov.in Ministry of Statistics and Programme Implementation (GoI) NSS 75th round	2017–2018	Difference in years with respect to other parameters
Expenditure: Household expenditure on services and durable goods	www.microdata.gov.in Household consumption expenditure	2014–2015	Limited components of durable goods and services
Land and Living space: Flat, independent house, floor area	www.mospi.gov.in Ministry of Statistics and Programme Implementation (GoI), NSS report no. 584	2017–2018	Type of dwelling and floor is the average of estimated number of households
Type of energy use for cooking rural and urban: Firewood chip and crop residue, LPG, dung cakes	www.mospi.gov.in MoSPI (GoI), *NSS report no. 584*	2017–2018	Percent distribution is presented in the data for the fuel used in cooking in rural and urban households
Transportation: Working of state transport undertakings	www.mospi.gov.in Motor vehicle statistical yearbook (India)	2014–2015	State operated transport limited to fleet operation
Food consumption: (Cereals, pulses, onions, sugar, Milk, butter, meat, sugar, jaggery)	www.mospi.gov.in www.indiastat. com Consumer Expenditure Survey	2011–2012 2006–2007	Per capita consumption expenditure Per capita consumption of selected foodstuff
Demographic information; urban-rural divide	www.mospi.gov.in MoSPI (GOI), NSS 75th round	2017–2018	Linking with other consumption parameters is not possible
Consumer products (e.g. Computer system, Electric purifier)	www.mospi.gov.in MoSPI (GoI) NSS 75th round	2017–2018	Percent of computer system and electric purifier is the percent of the number of households estimated in rural and urban areas and not the estimated numbers
Fuel consumption: LPG, motor spirit, kerosene oil, aviation turbine fuel, diesel, Naphtha	www.indiastat.com Petroleum product consumption in India	2017–2018	Specific to petroleum products
Electricity consumption	www.indiastat.com	2016–2017	State-wise annual electricity consumption per sq km per thousand population

Table 2.3 Percentage distribution of the socioeconomic variables in three zones in proportion to the total

Socioeconomic variables	Urban	Peri-urban	Rural	Total
Number of HH (% of total sample)	24.77%	22.49%	52.72%	1796
SC (%)	10.56%	10.14%	16.78%	13.75
ST (%)	11.46%	14.1%	14.36%	13.58
Average expenditure Rs per HH per annum	132649.3	87481.06	71401.24	90193.9115
Average total land holding Ha per HH	0.79	1.51	5.26	3.17
Average built-up space in sq.ft per HH	815.88	868.48	711.66	772.76
Average HH size	4.61	4.33	3.95	4.38
Total family members in total HH	1762	1752	4368	7882
Male (%)	52.92%	54.62%	53.64%	53.7%
Female (%)	47.07%	45.26%	46.28%	46.23%

Table 2.4 Percentage distribution of income classes in three zones in proportion to the total

Income group	Urban (%)	Peri-urban (%)	Rural (%)	Total (%)
Below 50 K	7.64	4.45	16.15	11.41
50 K to 1 Lakh	17.52	20.79	45.93	33.24
1–3 Lakh	24.26	38.86	28.51	29.78
3–5 Lakh	29.66	22.27	7.18	16.14
5–10 Lakh	17.07	12.62	2.11	8.18
10–30 Lakh	3.82	0.99	0.105	1.22

highest average expenditure per household per annum, but the rural zone holds the highest average total land holding whereas the built-up space per household is the highest for the peri-urban zones. The sample is representative data of seven states.

Table 2.4 shows the relationship between income groups and zones with their total percentage representation. Rural zone comprises majority of low-income category, i.e., below 50 thousand whereas about 3.82% in the urban zones represents the highest income group. The peri-urban zones comprise of middle-income zone within 1 lakh to 3 lakhs, which shows its high representation. The majority of sample falls under the 50 thousand to 1 lakh income group.

Focus Group Discussion (FGD), a perception-selection based questionnaire was used. The questionnaire was made to capture the thought process of the people, pertaining to their idea of carbon emissions. A series of activity-based discussion was administered with three different groups which was a part of a larger group and a fourth group where all the groups were combined to procure their collective opinion on the presented subjects. A detailed discussion followed which was recorded by means of devices and registration.

The following are the sections of questionnaire:

1. Food and consumables
2. Energy use
3. Transportation
4. Development and environment

2.3.2.6 Data Analysis

To develop an estimate for carbon emissions of the households from different zones, the consumption and expenditure data at unit level is very crucial. The availability of data based on consumption and emissions on the national portals is scarce. Within the study we have included 1950 households from nine agro-ecological zones. The assessment is solely based on the input from the residents where we enumerate the emissions from the consumption of total household energy, waste, transport, food and consumer products. For the expenditure standpoint carbon footprint estimations are based on health and medicine, education, clothes, textile and shoes, furniture, and other manufactured goods. The governing factors for the carbon footprint of a household have been divided broadly into zone and income categories. The study has employed a carbon assessment study wherein the household's expenditure and consumption has been integrated from the personal data through survey questionnaire and also an attempt has been made to compare the first-hand obtained data with nationally available data sets. The analysis does not include any carbon emissions related to government and industrial sectors.

Calculating the Footprint

The basis of calculating carbon footprinting involves emissions factor of the elements taken into consideration under consumption and expenditure heads. The reason for accounting carbon-based emissions for aforementioned heads is that the consumption of materials in terms of energy, food, travel, waste, and internet usage could substantiate upon the lifestyle of the people living in different zones of different states across the nation. The expenditure pattern can also be traced by examining the majority of expenses made by the households on particulars. Also considering the income bracket, we can discern a section of the income bracket, liable to make changes in their behaviour and adopt alternatives to reduce emissions. Conclusively, a comparison is made on the basis of average national emissions of household and per capita emissions. It is estimated that the consumption factor is responsible for 85% of emissions from the household in India; therefore, it could prove to be essential to estimate indirect emissions.

The emissions factors obtained from the Intergovernmental Panel on Climate Change (IPCC) and other official sources have been applied for the calculation of selected variable. The inputs from households were estimated in quantitative terms, and the units were defined in accordance with the type of variable measures in its own unit. The time period of units consumed for distinctive variables were estimated and converted in yearly terms by extrapolating the average use. Initially, the data was collected and analysed as given in Fig. 2.1. Categorization of sector-wise consumption and aggregate expenditure from the survey data was administered while covering the different product category was done which was linked with the NSSO data set for the comparison. As the required data was not entirely available, we took the data which was in consonance with primary data.

Series of analysis were conducted to collate a coherent output. The quantitative analysis presupposed collection of emission values. An excel-based HCF calculator was created to calculate carbon emissions from the consumption values of each

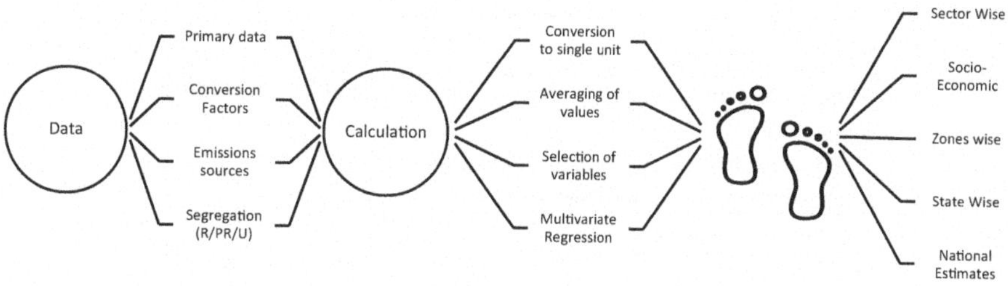

Fig. 2.4 Analytical process for the household carbon footprint (HCF) estimation

household. The conversion factors were used to convert all the units into a single metric (KgCO_2 equivalent) by equating with the consumption values obtained. Independent and dependent variables were determined by analysing cause-and-effect relationship, concomitantly, contingency tables were produced to analyse the relationship between multiple variables. Multivariate normal were introduced to carry out regression analysis to understand the correspondence between the procured and generated data sets. Variance Inflation Factor (VIF) was performed to test for any multicollinearity (Fig. 2.4).

Multiple Linear Regression

Description

Multiple linear regression is a statistical test used to analyse the relationship between dependent variable (response) and independent variables (predictor). The regression analysis provides predictors and the outcome variables which hold significance. The expression of the multiple linear regression equation as follows:

$$Y = \beta 0 + \beta 1X1 + \ldots + \beta nXn + \varepsilon i$$

Where Y is termed as dependent variable, β_0 is termed as intercept, β_1 is termed as slope parameter, X_n is termed as independent variables, and ε_i is termed as error term.

P value: *P* value is used to detect significance of the observed difference. The lower *p* value depicts greater statistical importance.

Coefficient: A regression coefficient is used to measure unit change in the variables holding predictor constant.

R Squared: Coefficient of determination R^2 is used to evaluate goodness of the fitted model.

$$R^2 = \frac{\text{SSR}}{\text{SST}} = 1 - \frac{\text{SSE}}{\text{SST}}$$

Where SSR = sum of squares due to regression, SSE= sum of squares due to error and SST= sum of square total.

When SSR is closed to SST, R^2 will, be closed to 1. That means regression explains most of the variability in Y and the fitted model is good. When SSE is

Table 2.5 Types of variables for the analysis in this study

S. No	Independent variables	Dependent variables (emissions)
1.	Household annual income	Biodegradable waste
2.	Expenditure	Non-biodegradable waste
3.	Total no. of family members in a household	Petrol
4.	Education qualification of the head	Electricity
5.	Occupation of the head	Biomass
6.	Type of house	LPG
7.	Land holding	Food
8.	Built-in space	Travel

closed to SST, R^2 will be closed to 0 that means regression does not explain much variability in Y and the fitted model is not good.

Data was first segregated zone wise (rural, peri-urban, and urban) and then state wise and combined. Variables used for multivariate regression analysis are as follows (see Tables 2.5 and 2.6):

2.3.2.7 Assessment of Landscape Dynamics

To understand the dynamics of landscape, the structural attributes (bio-physical properties) were analysed for two points of time (year 2000 and year 2020). The bio-physical classification was done with the help of supervised classification using maximum likelihood method in Google Earth Engine. The classification and regression tree (CART) model was used to classify the land use land cover in Google earth engine (GEE). CART is decision tree approach that was proposed by Breiman in 1984. The CART algorithm uses the Gini Coefficient as optimal test variances and segmentation technique that ultimately generates a binary tree-based decision tree classification. The Gini coefficient helps to classify the data on the basis of randomly selected training set. The CART has an advantage of easy manipulation and fast operational technique.

Band 4, 3, and 2 have been considered for FCC (False colour composite) of year 2000 in Landsat 5 TM and band 5, 4, and 3 have been considered for FCC of year 2020 in Landsat 8 satellite imagery for land use land cover classification. The CART method was applied to perform the land use land cover by collecting various training sets for each feature class. The training data sets are used to train the classifier and classify the feature collection. Furthermore, the layouts are prepared in ArcGis 10.8 software.

Table 2.6 Conversion factors for the units of various energy attributes

EA Unit of CF	Type of EA	Unit of EA	CF	Source
Consumption: energy				
$kgCO_2e/Kwh$	Electricity	Kilowatt Hour	0.85	MoEFCC (2010)
$kgCO_2e/L$	Petrol	Litres	2.296	
$kgCO_2e/L$	LPG	Litres	2.983	
$kgCO_2e/kg$	Biomass	Kilograms	1.747	
Consumption: food				
$gmsCO_2e/kg$	Cereal	Kilograms	119.5	Pathak et al. (2010)
$gmsCO_2e/kg$	Pulses	Kilograms	306.8	
$gmsCO_2e/L$	Milk	Litres	729.2	
$gmsCO_2e/L$	Fruits and vegetables	Kilograms	0.10	
$gmsCO_2e/g$	Chicken	Kilograms	846.5	
$gmsCO_2e/kg$	Fish	Kilograms	718.3	
$gmsCO_2e/kg$	Meat	Kilograms	12062.7	
Consumption: travel (work and non-work)				
$gmsCO_2e/km$	Train	Kilometres	41	India GHG Program (2015)
$gmsCO_2e/km$	Bus	Kilometres	105	
$gmsCO_2e/km$	Car	Kilometres	192	
$gmsCO_2e/km$	Motorbike	Kilometres	103	
$gmsCO_2e$/passenger km	Flight	Kilometres	0.17147	
Waste: type of waste				
$kgCO_2e/kg$	Biodegradable	Kilograms	2.538	Loyarte-Lopez et al. (2020)
$kgCO_2e/kg$	Plastic	Kilograms	0.534	
$kgCO_2e/kg$	Glass	Kilograms	0.395	
$kgCO_2e/kg$	Metal	Kilograms	0.883	
Internet: Data usage				
$kgCO_2e/GB$	Internet data	Gigabytes	0.046	Obringer et al. (2021)
Processed food: types				
$kgCO_2e/kg$	Potato chips	Kilograms	3.43	Pathak et al. (2010)
$kgCO_2e/kg$	Sauce	Kilograms	2.25	
$kgCO_2e/kg$	Bread	Kilograms	0.8	
$kgCO_2e/kg$	Sugar cookies	Kilograms	1.3	
$kgCO_2e/kg$	Milk chocolate	Kilograms	4.8	
Paper based material				
$kgCO_2e/kg$	Book	Kilograms	1.2	Boguski (2010)
$kgCO_2e/kg$	A4 sheet	Kilograms	0.004	
$kgCO_2e/kg$	Magazine	Kilograms	0.82	
$kgCO_2e/kg$	Daily newspaper	Kilograms	1	
Cleaning agents				
Kg CO_2e/kg	Soap bar	Kilograms	1.646	Wilson (2012)
Kg CO_2e/L	Shampoo	Litres	16.6	
Expenditure emissions				
Emission per 1000/-	Health and medicine	Metric tons	0.25	Parikh et al. (2009)
Emission per 1000/-	Education	Metric tons	0.02	
Emission per 1000/-	Clothes, textile, and shoes	Metric tons	0.04	
Emission per 1000/-	Furniture and other goods	Metric tons	0.03	
Emission per 1000/-	Recreational and cultural activities	Metric tons	0.03	

**EA* Emission attribute, **CF* conversion factor

3 Sectoral Contribution to Household Carbon Footprint (HCF)

The carbon footprint of households is a critical component of global greenhouse gas (GHG) emissions, impacting climate change significantly. Understanding the sectoral contributions to household carbon footprints is essential for developing effective strategies to reduce emissions. This overview explores the major sectors contributing to household carbon footprints, their impacts, and potential mitigation strategies. The energy sector is the largest contributor to household carbon footprints. This sector includes the consumption of electricity, heating, and cooking fuels. Transportation is another major sector contributing to household carbon footprints, encompassing personal vehicles, public transport, and air travel. The food sector contributes to household carbon footprints through the production, transportation, and consumption of food. The management of household waste, including solid waste and wastewater, contributes to GHG emissions, particularly methane from landfills and CO_2 from waste incineration. Each sector contributes differently to the overall emissions, and understanding these contributions is crucial for developing targeted mitigation strategies. By adopting renewable energy, enhancing energy efficiency, promoting sustainable transportation, encouraging dietary changes, and improving waste management practices, households can significantly reduce their carbon footprints. Collective efforts at the individual, community, and policy levels are essential to achieve substantial emission reductions and combat climate change effectively.

3.1 Household Carbon Footprint of India

The estimation of HCF based on survey of 13 locations across India has been done at two levels, one as basic HCF (Table 3.1) and the second as inclusive HCF (Table 3.2). Per capita carbon footprint from household consumption computed from primary consumption data of households varies from 1158 $kgCO_2e$ per year for IC-1 (<1 Lakh annual income) to 4768 $kgCO_2e$ per year for households in IC-6 (>30 Lakh annual income) resulting in a more than four-fold increase from the lowest income group to highest income group households. The value of footprint increases 14% from IC-1 to IC-2; however, a high leap in emission is noticed from IC-4 to IC-5, where the increase in emission is almost 40% from one income class to the next (Fig. 3.1).

S. Nautiyal et al., *Determinants and Stratification of Household Carbon Footprint*,
https://doi.org/10.1007/978-981-96-2739-4_3

Table 3.1 Composition of basic HCF estimated for India in $kgCO_2e$ per year

HCF composition	Food	Energy	Travel	Waste	Other non-durables	Total HCF	Per capita CF
Basic HCF-India average	672.28	2551.44	1283.65	974.41	142.96	5624.73	1272.56
% of total HCF	11.95	45.36	22.82	17.32	2.54	100.00	

Table 3.2 Composition of HCF (inclusive of all consumables) estimated for India in $kgCO_2e$ per year

HCF composition	Food	Energy	Travel	Waste	Others	Total HCF	Per capita CF
HCF (inclusive)-India average	916.23	2551.44	1283.65	974.41	779.50	6505.22	1471.77
% of total HCF	14.09	39.22	19.73	14.98	11.98	100.00	

Income inequality is not uncommon in India, but the gap is widening further. With higher economic growth, India has also been experiencing the higher level of inequality over the years. India's per capita gross domestic product (GDP) increased five times in last 2 decades (2000–2019). According to the World Bank estimate (using the World Bank Atlas method), India's GNI per capita (formerly GNP per capita) is $2120 in the year 2019. Data from one study by Oxfam revealed that, in 2019 the top 1% earned 21% of total income of India. This was 11% in 1990. The top 10% earned 56% of the total income of the country in 2019; the bottom 10% earned only 3.5%. The Oxfam survey also reported that India's richest 1% held 58% of the total wealth. IC-1 and IC-2 of the current survey together represents India's 50% of the population, whereas IC-6 represents <2% of India's population (Fig. 3.2). The corresponding HCF estimates for those income groups reveal that half of India's population is emitting <1350 $kgCO_2e$ per person per year and <2% is emitting 4768 $kgCO_2e$ per person per year.

Economic inequality can be calculated with reference to wealth, income, and consumption as the major variables. From the households' perspective, consumption data appears to be the most important factor as consumption reflects typical income levels. Consumption is considered to be a better measure of permanent income and is often less under-reported than current income. Consumption is an important tool in determining inequality and material well-being and is assumed to be more authentic than income or wealth which often fluctuates from time to time. This authenticity is because people tend to smooth their consumption over time. Distributional analysis in India is most commonly based on consumption data derived from India's National Sample Surveys. The validity of the consumption estimates depend on the type of consumption data is being measured. For example, consumption information related to utilities,

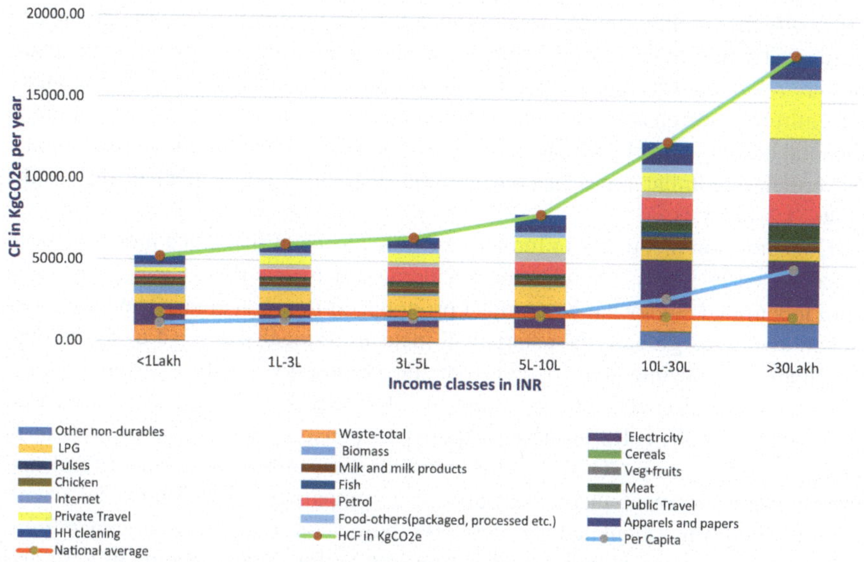

Fig. 3.1 Composition of all-inclusive household carbon footprint across economic classes
Note: Income groups (Household's annual income in INR) and population share in percentage: IC-1: <1 Lakh (43.44%); IC-2: 1–3 L (28.97%); IC-3: 3–5 L (15.70%); IC-4: 5–10 L (7.96%); IC-5: 10–30 L (2.27%); IC-6: >30 Lakh (1.63%)

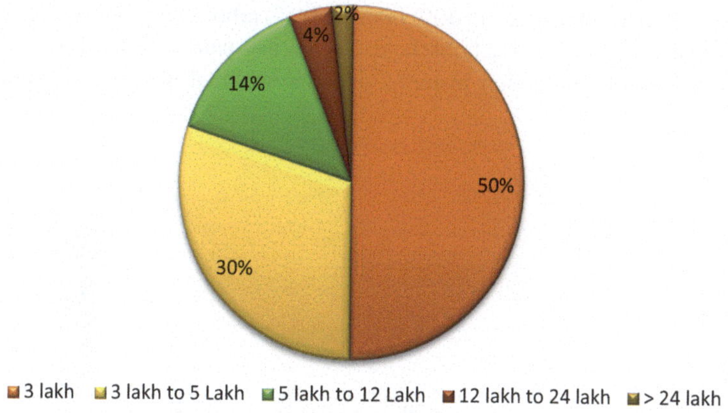

Fig. 3.2 Distribution of annual household income
Data Source: https://www.imrg.org/india-demographics/

groceries, and fuel tends to be accurate as compared to information on consumption on alcohol, clothing, and furniture.

Dearth of officially provided income data, many studies have used consumption expenditure data to measure the level of inequality in India. The low-income group spends a higher share of their income on food than the middle-income class and especially the wealthy, and the wealthy spends more relatively on entertainment and leisure activities. In terms of absolute value, the food-related consumption inequalities are relatively less than certain non-food items. The results of the current study are entirely based on actual quantity consumed under different items contributing to the total household carbon footprint which make the results closer to ground reality. Nevertheless, similar types of consumption disparities are evident in the results (Figs. 3.2 and 3.3).

Consumption inequality in India as a whole has been rising at a moderate pace since the early 1990s. The increase accelerated between 1993/4 and 2004/5 and then moderated somewhat being most pronounced in urban areas. The Gini coefficient also used globally is a statistical measure typically derived using income distribution data, intended to represent the income or wealth distribution of a nation. The Gini coefficient for India in 2018 was 47.9% which has increased from 34.4% in 2014. Rural-urban disparity in consumption is also prominent in higher income group (IC-5), but the emissions are almost uniform in rural and urban zones for lower income groups. The comparison of present results with carbon footprint calculated in 2003–04 based on consumption expenditure (Fig. 3.4) is showing significant elevation in CFs for all households in low- and middle-income households.

3.2 Contribution from Food Consumption

Generally, food accounts for 10–30% households' carbon footprint, where share is on higher side for low-income households. Types and pattern of food consumption have received global attention in terms of environmental impacts, nutrition, and

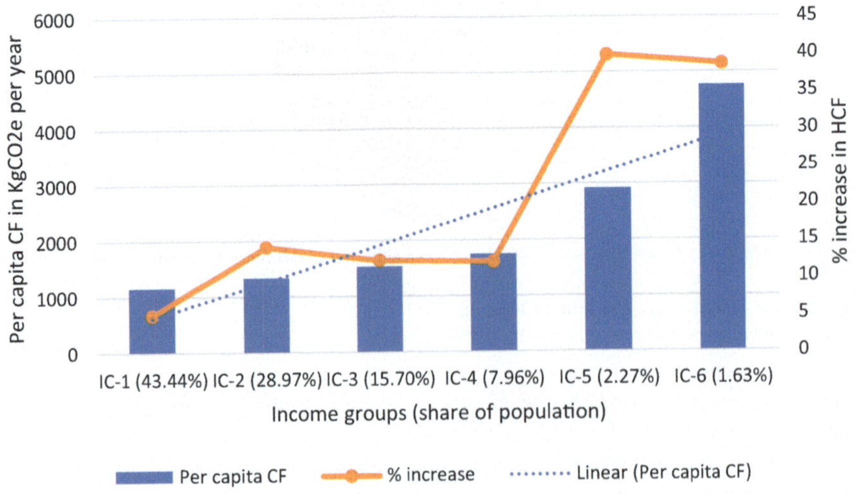

Fig. 3.3 Increasing trend of per capita carbon footprint along the increasing household income of Indian households

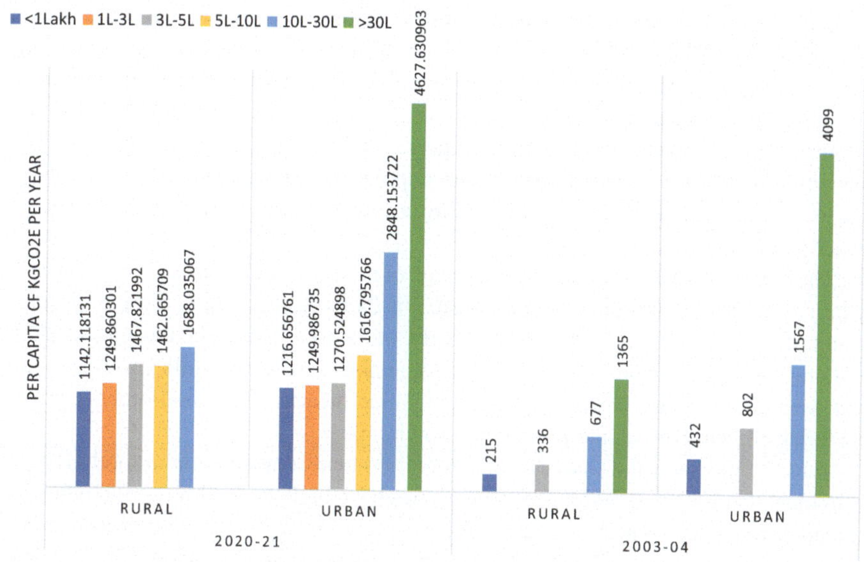

Fig. 3.4 Temporal variations of HCF across income groups in Rural and Urban India
Data source: Empirical survey (2020–21); Saluja and Yadav (2006), Parikh et al. (2009) (2003–04)

food security. The emission of GHGs from food items involves the entire value chain starting from production, processing, packaging, distribution, and consumption to waste disposal. On consumption part, numerous factors include social, economic, cultural, and agro-ecological parameters. The basic food items consumed in India are largely produced in India, except pulses and cooking oil where 17.5% and 59.3%, respectively, are imported (Green et al. 2018). This implies a comparatively lower carbon footprint of basic food required for attaining the nutritional security from balanced diet. Previous studies based on household expenditures reported that private transport, electricity use, and services demand in India accounts for a smaller proportion of individuals' carbon footprints, whereas food accounts for a slightly higher share as compared to developing countries (Lee et al. 2021). Indian diets reported the lowest carbon footprint scores in line with vegetarian diets in spite of including meat as well as the lowest energy intake per day (1739 kcal) (Gonzalez-Garcia et al. 2018). The emissions from dietary requirements based on gender, physical activity levels, and dietary pattern (veg–non-veg) have been estimated and revised several times where food carbon footprints of Indian dietary patterns are mostly aggregated from various expenditure heads. The current assessment shows 916.23 kgCO_2e per year per household which is lower than the GHG emission from balanced Indian diet which is approximately 1052 kgCO_2e per year per households (calculated based on Pathak et al. 2010). A balanced diet is one which contains different foods in quantities and proportion that the need for calories, minerals, vitamins, carbohydrate, fat, and other nutrients is met to withstand short duration of leanness (Gopalan 1985). Various research and dietary guidelines provide Indian balanced diet GHG emissions varying from 0.5 to 0.9 $kgCO_2$ per capita per day. These values are much lower than other developed countries (e.g. the United States: 3.83 kg; the Netherlands 2.86 kg and Germany: 2.25 kg). India's recommended diet footprint is even lower than global guideline, i.e., 1.36 kg (EAT-Lancet). Surprisingly, the study results reveal a majority of the population lagging behind in consuming the recommended diet, thereby producing a lower share in HCF with respect to other contributing items of HCF.

The basic food-HCF calculated as average of the entire 13 locations is 673 $kgCO_2e$ whereas processed and packaged food-HCF is 244 $kgCO_2e$; where these emissions account for 10.33% and 3.75%, respectively, of the total HCF. The basic food-HCF is contributed by cereals, pulses, fruits, vegetables, milk and milk products, and non-vegetarian group (fish, chicken, and other meat). The 'Food-others' group includes oil, tea, coffee, and processed food items.

Table 3.3 presents the shares of different food items contributing to food-HCF. It is evident that low- and middle-income groups have higher cereal footprint, whereas the emission share of non-vegetarian food is higher in higher income groups (>Rs. 10 Lakh annual income). The food footprint from the weakest economic class to the richest covered in the study has revealed an increase of more than threefolds. The food-HCF varies from 722.63 to 2208.68 $kgCO_2e$ from the lowest to highest income group (Fig. 3.5). The CF of processed and packaged food items vary from 215.18 to 602.29 $kgCO_2e$ per HH per year for income class-1 (<1 Lakh) to income class-6 (>30 Lakh). Figure 3.6 shows food-HCFs across income classes, where the income classes 1 and 2, i.e., the population less than Rs. 3 Lakh of household annual income are not achieving the emission corresponding to standard balanced diet.

Irrespective of heterogeneous random sampling across rural, urban, and peri-urban zones of 13 locations of India, the food footprint results of each location have shown tremendous variations which may be explained by agro-ecological factors and food production system, dietary habits and socio-cultural norms along with prime determinant, household income. Differences in food consumptions among the study locations and the three zones- rural, peri-urban, and urban are presented in Fig. 3.7. The lowest emission is 400 $kgCO_2e$ per HH per year in rural section of Maibang, a small town in Himalayan foothills of northeast India, whereas the

Table 3.3 Percentage share of different food groups in food-HCF ($kgCO_2e$ per year)

Percentage share of different food groups in food-HCF ($kgCO_2e$ per year)							
Income classes	Cereals	Pulses	Milk and milk products	Veg + fruits	Non-veg total	Food-others	Total food-HCF
<1 Lakh	9.07	7.51	28.05	4.56	22.63	27.01	100.00
1–3 Lakh	7.62	5.25	29.24	4.36	26.72	26.80	100.00
3–5 L	5.47	3.95	21.98	3.39	33.65	31.56	100.00
5–10 L	5.54	3.18	25.09	3.02	30.12	33.05	100.00
10–30 L	1.90	2.72	24.04	1.98	47.05	22.30	100.00
>30 Lakh	1.76	2.66	16.04	1.74	51.37	26.42	100.00

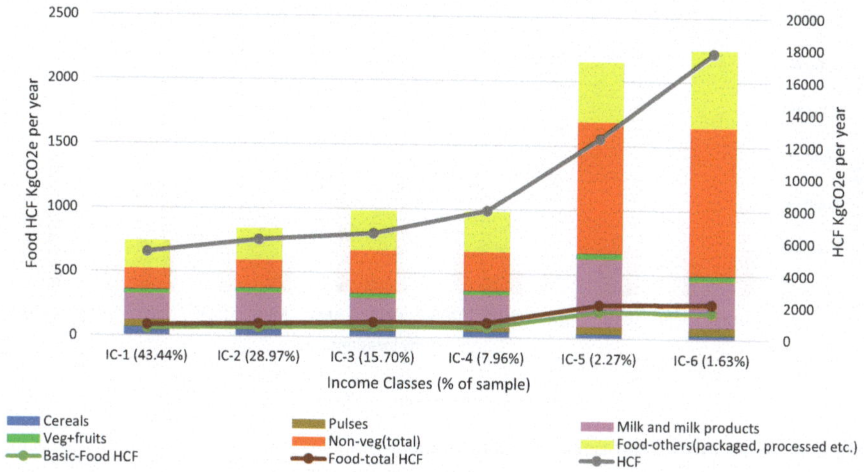

Fig. 3.5 HCF of food items across income classes across 13 study locations

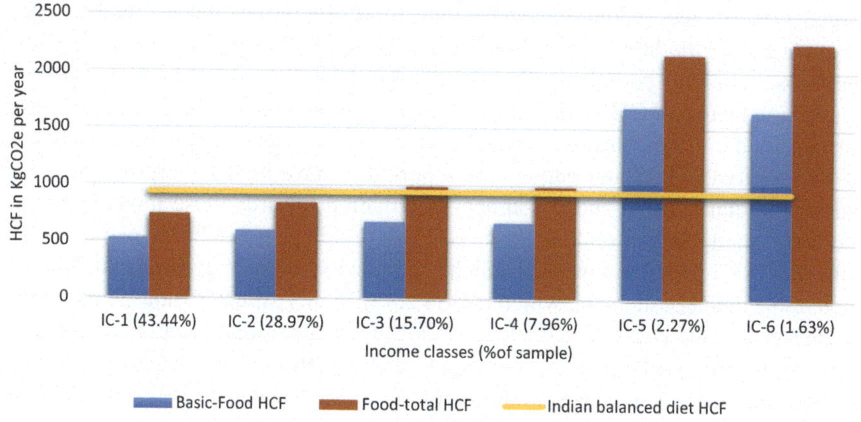

Fig. 3.6 Total and basic food-HCF across income classes with reference to carbon footprint of Indian balanced diet

highest food-HCF is reported from Bengaluru (urban) with a value of 2703 kgCO_2e per HH per year.

Among all the other food groups accounted in the estimation of Food-HCF, Tea/coffee has a significant contribution in the range of 149–258 kgCO_2e per HH per year (Table 3.4). Other smaller contributions include sauces/spreads, chips/namkeen, chocolates, confectionaries, etc. Frozen items have a contribution in the footprint only in selected households of high-income groups of Bengaluru and Noida. In India, meat is mostly consumed freshly cooked in spices (Table 3.5). Only 2% of the meat is processed in India and the survey results also show relatively low contribution from frozen non-vegetarian food even among meat-eating high-income

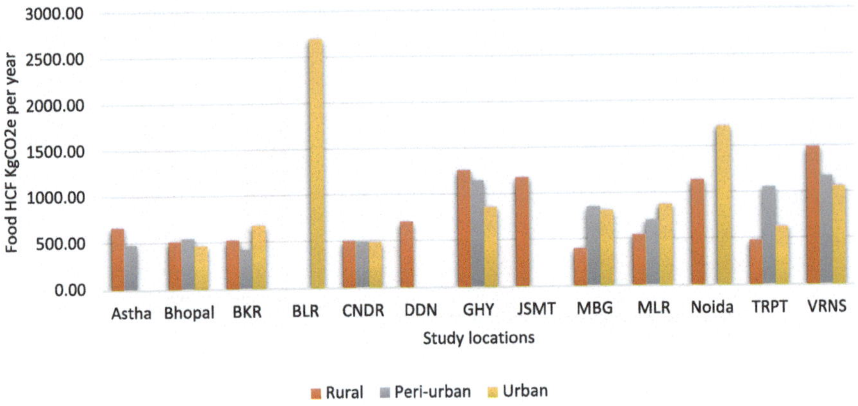

Fig. 3.7 HCF of food items across 13 study locations

groups. The increasing use of take-away food packaging associated with online meal deliveries is making the food sector's already massive carbon footprint even larger. Nevertheless, eating-out and online food orders have been found as very frequent among middle- and high-income urban as well as peri-urban households, the impact couldn't be holistically accounted in HCF because of lack of inclusive conversion factor for GHG estimation. The calculation done for these items are only for non-food consumables (e.g. packaging) and services involved in home delivery of cooked food. Pizza and burgers are reported to have higher footprints as compared to Indian cuisine and consumers' inclination towards those foods is mushrooming among Indian middle class. The India online food delivery market has a value of US$ 4.35 Billion in 2020 which is almost two times bigger than Australia's food home delivery segment, which is US$ 2.3 Billion. Estimates show 5600 tonnes of carbon dioxide-equivalent (CO_2-e) emissions in Australian food home delivery sector only from packaging (Arunan and Crawford 2021). With changing lifestyle, income and ease of ordering, more people have started not to cook in India and number is bound to increase (Fig. 3.8).

Transportation of food from farm to plate, i.e., the "food miles" is a worldwide concern in terms of climate change. Being most of the food items produced domestically (Table 3.6), India has been avoiding that part of carbon footprint. The survey results are promising in that respect as 23–24% of the studied population has been consuming self-grown cereals and milk/milk products to some extent. Fruits, vegetables, and non-vegetarian foods consumed by the respondents' households are largely locally grown (90–97%) across 11 study locations, except in half of the households in Noida and Bengaluru.

Table 3.4 $KgCo_2e$ emissions from other food groups per household per year

Income classes	Alcohol	Tea/coffee	Online food	Eating out	Sauce spread	Chips/namkeen	Chocolate	Bakery and allied	Others including frozen	Total
<1 L	11.60	149.23	0.29	0.83	4.25	9.23	10.2	9.56	31.59	215.18
1–3 L	12.90	159.37	0.28	0.71	4.6	11.84	14.82	14.11	38.89	244.62
3–5 L	9.92	218.36	1.87	3.45	5.37	12.8	14.4	15.03	40.90	312.18
5–10 L	9.20	189.2	2.12	8.14	8.39	25.84	20.06	19.55	42.20	315.50
>10 Lakh	7.10	258.42	13.1	12.38	13.93	27.78	74.59	17.99	184.10	602.29

Table 3.5 Domestic supply and imports of major food groups in India

Food items	Domestic supply (1000 MT) Total	Import share (%)
Cereals	21652	0.1
Starchy roots	5356	0.2
Sugar crops	34120	0.0
Sugar and sweeteners	3258	4.1
Pulses	2169	17.5
Tree nuts	198	53.9
Oil crops	4977	0.29
Vegetable oils	1882	59.3
Vegetables	11853	0.1
Fruits	8227	1.0
Spices	347	3.3
Meat	463	0.0
Animal fats	403	0.1
Eggs	376	0.0
Milk	13344	0.0
Seafood	699	0.7
Aquatic products	0.4	0.0

Data source: http://faostat3.fao.org/home/E

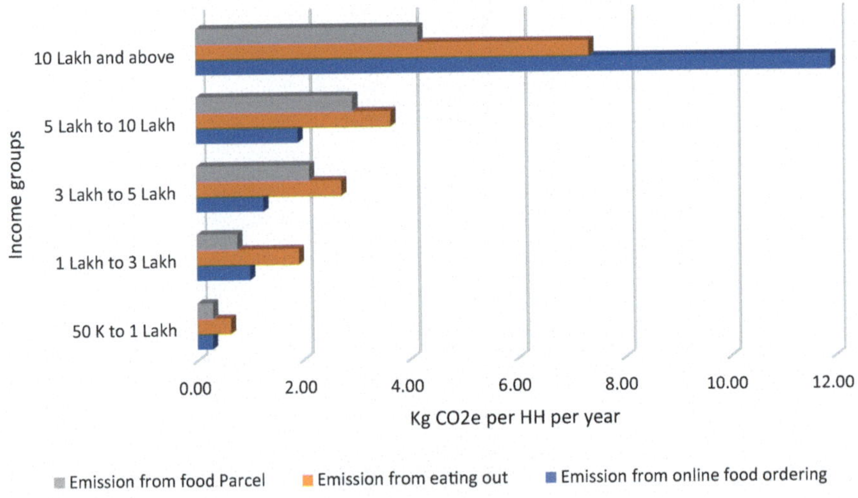

Fig. 3.8 Emissions from non-home cooked food habits

3.3 Contribution from Energy

The contribution of the energy emission accounts for the major portion of household emissions which can be over 50% of Indian households. Lighting, heating, and cooking are some of the activities which utilize most of the energy. On account of

Table 3.6 Percentage of households sourcing different food items

Source of food	Cereals	Pulses	Milk and milk products	Oil	Fruits and veggies	Non-veg
Self-grown	23.45	9.92	24.74	1.86	5.59	1.23
Local shop	68.24	82.02	71.36	91.23	90.92	96.93
Super market	7.48	7.24	3.35	6.08	3.15	1.23
Online	0.83	0.82	0.55	0.82	0.33	0.61

Table 3.7 Percentage share of different energy groups in energy HCF (kgCO_2e per year)

Income group	Petrol	Electricity	LPG	Biomass	Total
<50 k	3.79	32.46	27.91	35.84	100.00
50 k–1 L	6.54	51.64	22.98	18.85	100.00
1–3 Lakh	17.45	46.51	27.90	8.13	100.00
3–5 L	31.59	32.40	31.24	4.77	100.00
5–10 L	23.39	40.67	34.35	1.59	100.00
10–30 L	27.76	59.14	12.90	0.20	100.00
>30 Lakh	35.27	54.70	10.03	0.00	100.00

differentiated zones and disparity in income classes, variations in the use of residential energy are evident from the study. When it comes to accessibility to energy sources the most of the developed areas have grid electricity, LPG, and other liquid fuels whereas underdeveloped areas have high dependency on the traditional forms of resources. The fuel type and the quantity of consumption are responsible for carbon emissions which have adverse effects on the personal and global health. The emissions due to the use of fossil fuels in cooking, running vehicles, heating, etc., account. One observation provided insight that dependency on one type of energy source has larger implication on the carbon emissions when compared to the availability of alternative energy sources. The present research put forth the perspective of the portion of energy use by different income groups.

The HCF composition, from the analysis provides an overall view where the average HCF of energy alone forms the majority of 39.22% of the total household emissions of all the locations. Table 3.7 represents the percentage of types of energy use in households; we have considered petrol as a source of energy because some of the respondents have admitted using them for heating and cooking, mostly in the peri-urban and rural areas. From the table we can discern the highest and the lowest energy emissions contribution of the income classes. An increasing trend can be observed for most of the energy-type from low- to high-income classes excluding biomass. The highest emitter from the petrol and electricity belongs to 30 lakhs and above income class, whereas the income class 5–10 lakhs has highest LPG emissions and below 50 k income class emitting carbon from the use of biomass.

The selected study areas from the following states Assam, Jharkhand, Karnataka, Madhya Pradesh, NCR, Tamil Nadu, Uttar Pradesh, and Uttarakhand, by and large, fall under different climatic zones. The inferred data observes disparity between the

Table 3.8 Emissions from all the study locations across different sectors

Region	Energy	Food	Waste	Travel	Internet
Ashta	3133.18	513.95	1240.36	168.95	1.79
Bhopal	1587.97	372.49	653.64	1123.60	13.86
BKR	1126.16	446.43	586.36	1352.47	40.91
BLR	3337.39	2079.29	1575.91	2320.08	114.84
CNDR	815.00	378.07	628.66	1388.72	14.34
DDN	4542.07	542.31	673.49	864.62	37.36
GHY	817.59	838.75	1194.10	904.61	24.18
JSMT	3636.76	907.72	1116.77	726.96	27.91
MBG	1089.63	527.96	970.47	214.07	0.00
MLR	2548.81	542.41	1664.29	2577.58	38.63
Noida	3909.82	1100.34	1677.40	3266.44	40.04
TRPT	3211.96	553.57	160.58	2119.03	53.36
VRNS	3937.61	955.21	1186.84	1177.57	74.88

consumption and emissions between these areas (Table 3.8). In terms of energy, the Western Himalayan population depends on natural resources in majority, the consumption of biomass is more when compared to other areas because it constitutes animal residue and conventional cattle rearing is still prevalent among the natives of the Western Himalayan region, fuelwood and the stubble used as fuel is also a factor. The Western Himalayan population has less appliance usage when compared to urbanized areas; therefore, it has less emission. The Upper/Trans Gangetic plains and Southern Plateau have significantly higher emissions when compared to any other agro-climatic zones; this can be attributed to the development and urbanization of the area. The dense population is also one of the reasons for high emissions value of these areas. The Central Plateau, Middle Gangetic Plains along with Eastern Himalayas have shown moderate elevation in emissions with an increase in amenities, income, and standard of living. Though the energy consumption-emissions are higher for the Upper/Trans Himalayas and Southern Plateau, some parts of Western Himalayas and West Coast Plains will shortly emulate the household emissions of the Upper Gangetic and Southern Plateau. Distinctively, the emissions from food consumption are seen higher in the Upper/Trans Gangetic Plains and lowest in the Western Himalayas, which includes cereals, pulses, veg and non-veg food items. The trend in the food consumption-emissions can be observed when we move from north to southward direction where the highest emitters fall under Upper/Trans Gangetic Plain followed by Middle Gangetic Plain, West Coast Plains and Hills, and Southern Plateau and Hills. Moving from east to westward direction, the residents of the Eastern Himalayas are the highest emitters followed by Eastern Plateau and Hills, Central Plateau and Hills, and Western Plateau and Hills as per the outcome obtained. As far as the emissions from waste disposal is concerned, the residents of Southern Plateau and Hills are the lowest emitters which is justified by their equitable waste management practices, whereas the Upper/Trans stands as the highest emitter in the waste emissions category. Emission from both public and private transportation is high in the Upper Gangetic Plains, Middle Gangetic Plains as well as Southern Plateau and Hills as some of the states within these agro-climatic zones encompasses highest percentage of National Highways and State Highways in the country and are also essential economic zones in India. They are having highest number of vehicular possession when compared to other regions. The lowest emitters in terms of transportation are the Western Himalayan region. High urbanized areas with high population density provide insight into the increased emissions through internet data usage of the Middle Gangetic Plains, Upper/Trans Gangetic Plains, and the Southern Plateau and Hills. The emissions from the expenditure forms a substantial portion of the total HCF. The highest expenditure-emissions estimations are from the Southern Plateau and Hills, Upper/ Trans Gangetic Plains as states under these regions holds better health, education, textile infrastructure and provides range of goods and services and the households have sound financial backup than other regions.

The graph provides an overview of the emissions of all the selected study areas under various agro-climatic zones concerning energy use, food, energy, travel, waste, and internet use (Fig. 3.9).

Figure 3.10 and Table 3.9 provide total and per capita emissions pertaining to the type of agro-climatic zones. The Southern Plateau and Hills with Upper/Trans Gangetic Plains have both high emission in total and the per capita emissions in comparison to other states, whereas the emissions from the household of the Eastern Himalayas and Eastern Plateau and Hills account for the lowest total and per capita emissions as they operate in moderation of both renewable and non-renewable source of energy. There is also less variation in the consumption as they tend to save more use less, and the relation is monotonic between wealth increment and availability of resources for increased consumption.

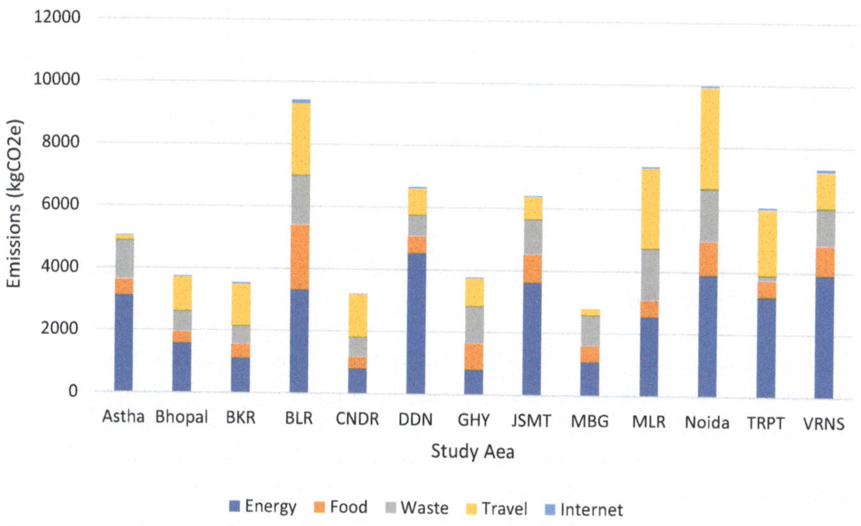

Fig. 3.9 Sectoral emissions and contribution to HCF in all the study locations

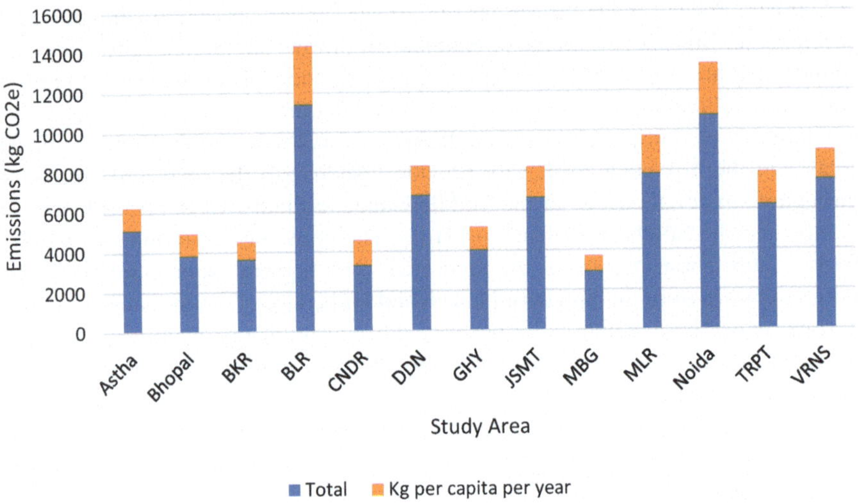

Fig. 3.10 Total HCF and per capita emissions across all the study locations

Table 3.9 Total household emissions and per capita emissions from all the areas combined

Region	Total	Kg per capita per year
Ashta	5139.71	1129.07
Bhopal	3870.30	1086.38
BKR	3679.78	870.82
BLR	11447.49	2905.41
CNDR	3341.70	1246.68
DDN	6837.12	1462.37
GHY	4084.13	1123.23
JSMT	6697.42	1536.04
MBG	2970.56	753.24
MLR	7864.16	1868.44
Noida	10778.68	2578.04
TRPT	6266.92	1648.21
VRNS	7560.06	1442.09

3.4 Discussion

The Intergovernmental Panel on Climate Change Special Report on 1.5 °C (IPCC SR1.5) states that lack of access to clean and affordable energy for cooking is a major policy concern in many countries where major parts of the population still rely primarily on solid fuels for cooking (Roy et al. 2018). The International Energy Outlook Report, 2017 reported that in 2030, 580 million people will depend on solid fuels for cooking in India. Fuel burning is estimated to add the equivalent of another 8–16% of the global warming caused by CO_2 (Bailis et al. 2015). It is also important to recognize the complexity associated with accounting in terms of the time it takes for replacement trees to sequester the CO_2 emitted by burning a felled tree, and the rate of change in CO_2 sequestration as trees mature. Clean cooking solutions address the most basic needs of the poor. Furthermore, reducing black carbon, methane, and other short-lived climate pollutants would have substantial co-benefits on health and local air quality but can in the short term contribute significantly to limiting global warming to 2 °C and 1.5 °C (de Coninck et al. 2018; Batchelor et al. 2019). A crucial point which should be prioritized in policy debate is how these cleaner alternatives diffuse and get adopted by mass with affordable price, which are also proven for reliability and long-term sustainability.

4 Rural–Urban Disparity and Prospects for Emission Reduction

The rural–urban disparity in carbon emissions is a significant aspect of global environmental challenges. This disparity stems from differences in lifestyle, economic activities, energy consumption patterns, and access to technologies between rural and urban areas. Understanding these differences is crucial for developing targeted strategies for emission reduction. Urban areas typically exhibit higher per capita energy consumption compared to rural areas. This is due to the concentration of industrial activities, higher standards of living, and greater access to energy-intensive services and amenities in cities. Urban households often use more electricity for lighting, heating, cooling, and appliances. In contrast, rural areas generally rely more on traditional biomass and less efficient energy sources for cooking and heating. This results in lower overall energy consumption but higher emissions of pollutants and less efficient energy use.

4.1 Zonal Disparity

It was inferred from the data that the average total emissions estimation was 672.28 kgCO_2e, merging all types of food taken into consideration (Table 4.1). Animal-based food products emit more than plant-based food types and together they constitute 10.33% of total emissions (Fig. 4.1). Agriculture-based food materials have high emissions in rural areas as land use, irrigation, and tilling of soil with other factors are crucial for the growth cereals, pulses, and vegetables in addition to consumption at source. The emissions from animal-based food products are high as the factors, such as accessibility, affordability, and supply chain, add to the consumption of meat-based food products in urban zones and peri-urban zones. The packed and processed food products which are also a part of total food consumption is responsible for 3.75% of the total average emissions of the households.

Emissions from the sources of energy (petrol, electricity, LPG, and biomass) combined can be observed in Fig. 4.2, and the average energy emissions of all the

S. Nautiyal et al., *Determinants and Stratification of Household Carbon Footprint*,
https://doi.org/10.1007/978-981-96-2739-4_4

Table 4.1 Total emissions in $KgCO_2e$ per year from various sources in households (Urban, Peri-urban and Rural zones)

Zone	Food	Energy	Travel	Waste	Others	HCF	Per capita CF
Urban	791.2883	2681.254	1768.92	992.5671	292.448	6142.925	1492.366
Peri-urban	626.418	1976.398	1577.061	846.942	145.9936	5172.812	1196.648
Rural	630.7129	2297.701	899.3159	1032.984	128.1147	5372.382	1179.616

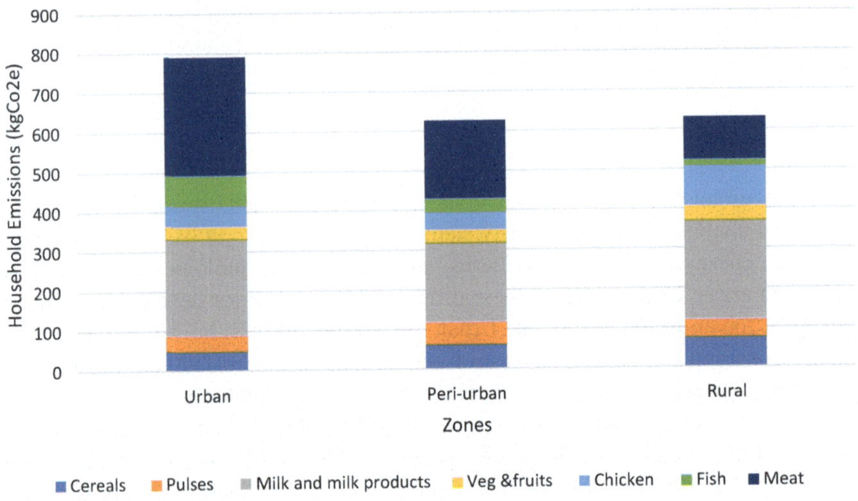

Fig. 4.1 Difference in the zonal emissions-Food

locations is estimated to be 2551.44 $kgCO_2e$ which forms the major share of the total emissions of all the locations combined, i.e., 39.22%. The emissions from the households are highest in the urban zone (3043.01 $kgCO_2e$) which suggest the dependency on the non-renewable sources of energy. The rural households' emissions (3025.10 $kgCO_2e$) are also high as they are more dependent on traditional form of energy sources; hence, the emissions from biomass energy use are high as they have cattle, crop residue, and access to fuelwood and other solid fuels which fulfils their cooking and other needs, whereas in urban and peri-urban areas the use of non-solid fuels has less emissions compared to the use of biomass in the rural setting. Reasons for higher electricity emissions in rural area is the size of family, as we know that in rural areas average members of household (4.55) are higher because they prefer to live in a joint family system. Though the use of appliances is higher in urban areas, the use of machinery for farming purposes, for instance, motors to draw water and irrigation system uses both liquid fuel and electric energy, and the frequency of use of the electricity-run appliance, though less in number than urban households, are higher in rural households. It has been also observed that the supply of electricity is equitable in some areas for the three zones, but the smart consumption and efficient use of equipment has resulted in reduced emissions for the urban setting.

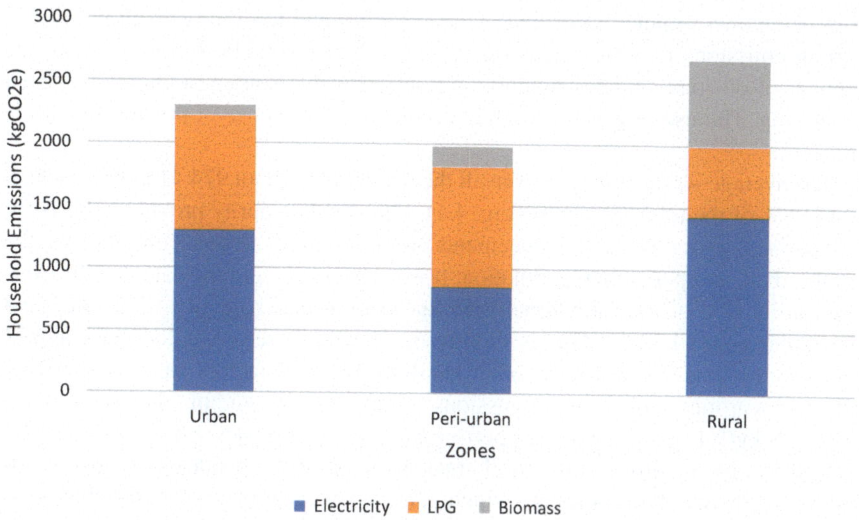

Fig. 4.2 Difference in the zonal emissions-Energy

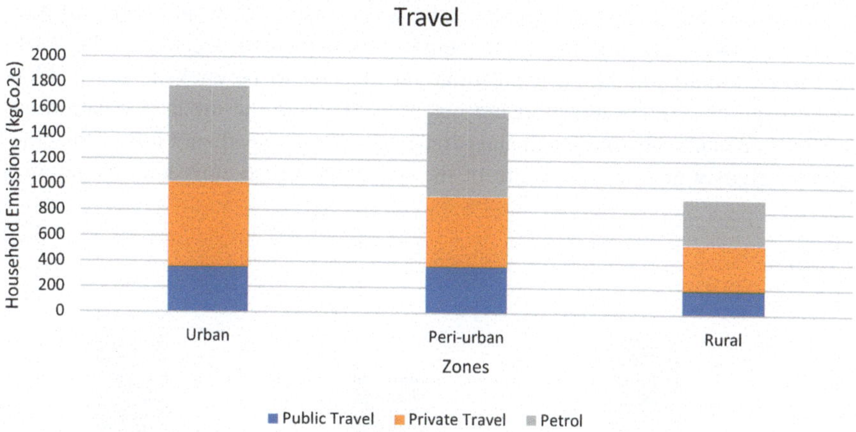

Fig. 4.3 Difference in the zonal emissions-Travel

4.2 Emissions from Transportation and Travel

The emissions from transportation are shown in the figure (Fig. 4.3) provides emissions from public and private use of vehicles which constitutes second largest portion of the total emission, at 19.735 which is on average, 1283.65 kgCO_2e. The emissions from transportation generally depend upon distance travelled and burning of fossil fuels in the vehicles. As we know most of the economic activities takes place in urban settings and people prefer to commute to office and other motorized trips using own vehicles according to their comfort and feasibility result in spike in carbon emissions of urban areas for private vehicles (663.06 $kgCO_2e$). Also, the number of on-road vehicles running on non-renewable source of energy has been increasing. The use of petrol is high in urban areas as the use of private vehicles is frequent.

The average waste emission from all the locations is about 974.41 $kgCO_2e$ which is 14.98% of the total emissions (Fig. 4.4). The zonal disparity provides a basis for comparison, rationale, and inducements for household carbon emissions of the rural, urban, and peri-urban population. It was observed from the data that the average family size of the rural zone (4.55 $kgCO_2e$), urban zone (4.11 $kgCO_2e$), and peri-urban zone (4.32 $kgCO_2e$) is slightly significant; therefore, it does have impact on the emissions. Furthermore, other ancillary lifestyle factors have conclusively affected carbon emissions. Emissions from the waste in the rural zone (1032.98 $kgCO_2e$) is higher than peri-urban and urban zones which is due to the lack of storage facilities in the rural areas, as most of the residents are directly or indirectly involved in farming, which makes sustainability of perishable food impossible, hence the wastage. Inherently half of the consumables and food materials are wasted just after the harvest because of the inadequacy of robust storage infrastructure. Another reason is, if the produce does not match the standard quality and shape the produce is disregarded. The waste emissions from urban and peri-urban are close. There reason for the emission in peri-urban areas could be attributed to customary practices, overestimation of food to be cooked, and reckless waste disposal of non-biodegradable materials. Though the urban population has better waste management system and is fairly educated, they do not follow the practices religiously and are not aware of the impact of waste emissions. Sometimes

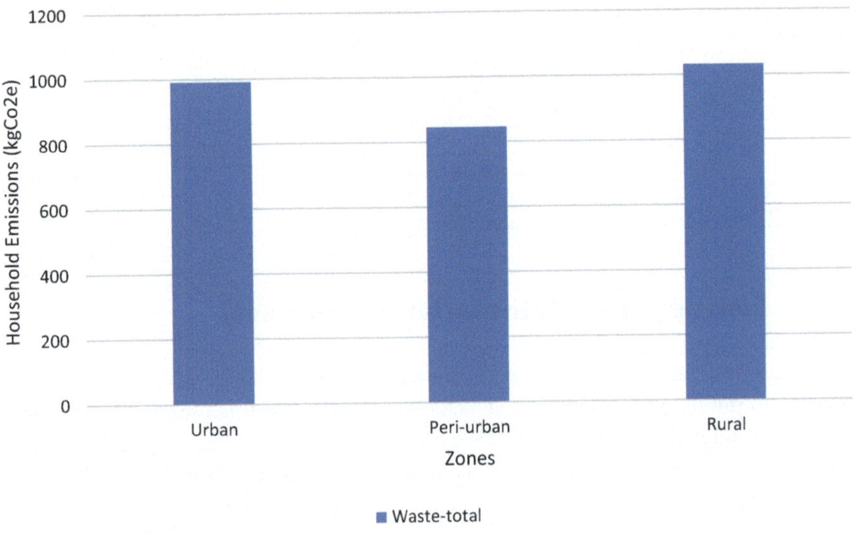

Fig. 4.4 Difference in the zonal emissions-Waste

they are not sure of the quantity of materials to be bought and do not check their inventories.

The digital environment has enhanced the dependency on internet and data-enabled devices (Fig. 4.5). Electricity is the major energy source in the transmission of data. The emission from urban areas is higher (43.72 kgCO_2e) as the use of internet-based services, use of social media, and the nature of work in the urban setting require use of internet. The average for the total internet usage across all the locations is 33.55 kgCO_2e and is 0.5%of the total emissions.

The non-durables form accounts for 2.2% of the total emissions with an average of 142.96 kgCO_2e. Use of consumption of non-durables are also high in urban households as the range of products such as soaps, detergent, and paper products are highly accessible and frequent purchase of such non-durables are high in the urban households followed by peri-urban, and rural which are almost at par with each other. The economic impact on non-durable is not governed by market situation and it certainly falls under necessary goods. The appeals and household cleaning agents are responsible for 4.21% and 5.07% of the total emission calculated for all the areas, with an average of 274.04 kgCO_2 and 329.96 kgCO_2 yearly.

The total disparity in the kgCO_2e per year between the three zones provides an insight into the urban emission (6383.85 kgCO_2e) which is the highest, attributing it to developmental activities and high consumption of resources (Fig. 4.6). The average total HCF estimates up to 6505.22 and average per capita emissions up to 1471.77 kgCO_2e which when compared suggests that per capita emissions are also high for urban households (1656.72 kgCO_2e) followed by peri-urban (1331.07 kgCO_2e) and rural (1275.69 kgCO_2e) which proves that the consumption of energy, namely electricity, petrol, animal-based products, and use of private vehicle are the major factors responsible for most of the emissions in urban households.

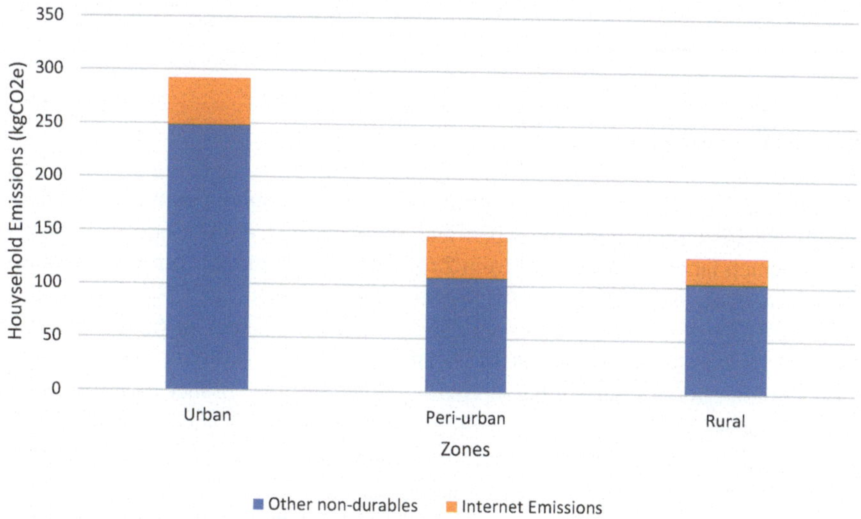

Fig. 4.5 Difference in the zonal emissions-Internet

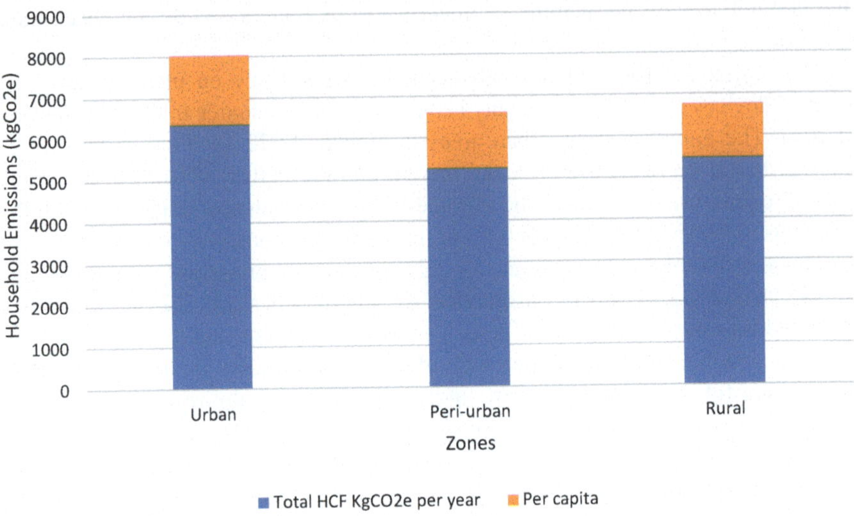

Fig. 4.6 Difference in the zonal emissions-total per capita

An increase in income also results in increased amenities, gadgets, and devices which in turn entail more energy consumption in the urban households though the usage is not that frequent. Households with more than average number of members, as in the case of rural households, lead to increased carbon emissions with energy use, for example, an increase in the use of biomass in rural households with every additional member. Biomass is a low-cost source of energy and is easily available. The urban and peri-urban houses mostly use LPG which emits less carbon when compared to biomass. The distribution in the given graph suggests that the per capita emissions of urban total are 25.95%, whereas it is roughly 25.20% for peri-urban and 23.24% for rural zone.

4.3 Discussion

Research on household carbon footprints (CFs) in developing countries highlights significant disparities between urban and rural areas. Urban residents typically exhibit much larger CFs, ranging from 2 to 9 times those of their rural counterparts (Connolly et al. 2022). In India, for instance, the average urban CF is around 2.33 tonnes of CO_2 equivalent (CO_2e) per capita annually, whereas in rural areas, it is about 0.56 tonnes (Bera et al. 2024). Kaul et al. (2023) supports these findings, reporting that urban CFs can be approximately eight times higher than rural CFs. Several factors contribute to these differences, including higher electricity consumption, more extensive transportation use, and overall lifestyle choices in urban settings (Connolly et al. 2022; Bera et al. 2024).

Moreover, over time, carbon inequality has been on the rise, particularly in rural regions. Sri and Banerjee (2023) note that the rural carbon Gini coefficient, a measure of inequality, increased from 0.308 to 0.324 between 2003–04 and 2015–16. This widening gap is largely attributed to evolving consumption patterns, notably the growing transportation needs in rural areas (Sri and Banerjee 2023). These studies collectively underscore the growing divide in carbon footprints between urban and rural households, driven by different consumption habits and lifestyle choices, with urban areas consistently showing higher carbon emissions.

The gap between rural and urban areas is a longstanding issue that spans economic development, infrastructure, service access, and environmental impacts. Urban regions typically have higher population densities, better infrastructure, more economic opportunities, and greater access to services. In contrast, rural areas often struggle with lower incomes, limited access to essential services, and a dependence on agriculture and natural resources. Environmentally, these disparities shape patterns and levels of greenhouse gas (GHG) emissions. Urban centres are significant contributors to global emissions due to industrial activities, high energy consumption, and transportation demands, accounting for about 70% of global CO_2 emissions. However, cities are better equipped with technologies and policies to reduce emissions, such as public transportation systems, renewable energy infrastructure, and energy-efficient buildings. On the other hand, rural areas typically have lower per capita emissions but face distinct challenges. The widespread use of biomass for cooking and heating in many rural regions contributes to deforestation and indoor air pollution, impacting the environment and public health. Additionally, rural economies, heavily reliant on agriculture, are vulnerable to climate change and often lack the resources needed for adaptation and mitigation.

Urban households have higher energy demands than rural households due to greater consumption levels and access to energy-intensive goods and services in urban areas (Lenzen et al. 2004). A significant disparity in energy consumption between urban and rural households is driven by income differences, access to modern energy sources, and lifestyle variations (Pachauri 2008). It is shown that urbanization significantly increases household energy consumption and material footprints, with urban households demonstrating much higher energy use and environmental impact than their rural counterparts (Wiedenhofer et al. 2011) primarily due to their greater use of electricity, modern fuels, and higher income-driven consumption patterns (Liu et al. 2009). There are significant differences in energy demand between urban and rural households, with urban households consuming more energy and having greater environmental impacts due to the increased use of modern appliances, heating, and cooling systems. It calls for targeted policies to address the challenges posed by urbanization (Zha et al. 2017). Addressing the significant disparities in emissions across Indian states necessitates region-specific strategies, such as enhancing industrial efficiency, adopting renewable energy, and improving urban planning in high-emission states, while states with high carbon storage capacities should focus on forest conservation and sustainable land management to maximize natural carbon sequestration (Ramachandra 2012).

Indian states like Bihar, Uttar Pradesh, and Madhya Pradesh are struggling with significant poverty; it should focus on integrating poverty alleviation with environmental sustainability, creating green job opportunities, and improving access to clean energy in rural areas to achieve targeted and sustainable poverty reduction (NITI Aayog 2021). The importance of balanced development across Indian states by addressing disparities in carbon footprints and economic growth, recommending the promotion of low-carbon economic activities, investment in renewable energy, the development of policies aligning economic growth with environmental sustainability, the implementation of carbon capture and afforestation projects, and strengthening climate resilience in vulnerable, impoverished districts (Nandy 2023). Supporting the poorest half of the global population, who contribute minimally to global emissions but are disproportionately affected by climate change, is vital. This support should include providing financial and technical assistance for adaptation, ensuring equitable access to clean energy and sustainable technologies, and promoting sustainable agricultural practices to enhance resilience and reduce emissions (Gore 2015).

Rural–urban disparities present both challenges and opportunities for reducing emissions. Urban areas, while being the largest sources of emissions, also have greater potential for rapid and large-scale reductions. Rural areas, though contributing fewer emissions, require specific, targeted solutions. By focusing on decentralized renewable energy, sustainable agriculture, improved transportation, supportive policies, education, and cross-sector collaboration, emissions can be reduced while addressing socioeconomic inequalities between rural and urban regions. These efforts are crucial for achieving a fair and sustainable transition to a low-carbon future.

5 Behaviour, Perceptions, and Knowledge of Households

Climate change, an unpredicted global challenge, stands at the intersection of environmental, social, and economic dynamics. The crisis lies in the intricate relationship between carbon emissions, individual and household behaviours, and their profound impact on the planet. The global climate is undergoing significant transformation primarily attributed to anthropogenic activities, notably the excessive release of greenhouse gases (GHGs) in the atmosphere. Carbon dioxide (CO_2), methane (CH_4), and nitrous oxide (N_2O) are the major contributors trapping heat and intensifying the greenhouse effect leading to global warming. The consequences are evident in rising sea levels, extreme weather events, and disruptions to ecosystems. Carbon emissions primarily result from the burning of fossil fuels and deforestation, which are major contributors to GHGs. Industrial activities, transportation, and energy production account for a substantial share escalating the levels of atmospheric carbon. Understanding and mitigating carbon emissions are pivotal in curbing the progression of climate change.

On an individual and household level, consumption patterns, energy choices, and daily practices play a crucial role in carbon footprint. Choices regarding transportation modes, energy efficient appliances, waste management, and dietary habits contribute significantly to the overall carbon footprint. A paradigm shifts in behaviour change is essential for sustainable lifestyle. Several international agreements and initiatives like Paris Agreement, aimed to unite nations in the fight against climate change. Commitments to reduce emissions, transition to renewable energy, and enhance climate resilience are integral components of these efforts. Global cooperation is imperative to tackle climate change collectively. India with burgeoning population and rapid economic growth faces distinctive challenges concerning climate change. Balancing development goals with environmental sustainability is a delicate task. The country is working towards renewable energy adoption, afforestation, and sustainable urban planning to address its carbon emissions. In Indian context, the impact of climate change is intimately tied to cultural, economic, and social aspects. Household behaviour influenced by traditions, and socioeconomic factors

S. Nautiyal et al., *Determinants and Stratification of Household Carbon Footprint*,
https://doi.org/10.1007/978-981-96-2739-4_5

shapes the carbon footprint in unique ways. Addressing these factors is vital to address the climate mitigation strategies. Climate change, carbon emissions, and individual household behaviour are crucial for devising effective strategies to mitigate the crisis. The transition towards a sustainable future recognizes collective efforts a commitment to balance development with environmental stewardship.

5.1 Behaviour, Perceptions, and Knowledge of Households

Household actions and government policies are both necessary to mitigate the climate change. Households serve as the key units of consumption which are affected by individual as well as collaborative behaviour of the members of the households are agents of change increasingly become a crucial element. The behaviour of households has to be understood in designing transition to low carbon alternatives to facilitate a bottom-up mechanism for strategizing carbon emission reduction (Niamir et al. 2020). In this study, the comprehensive behavioural analysis section is not only providing numbers that would help in decision-making but also developing narratives of household knowledge, willingness, and motivation for climate change mitigation.

Perceptions, behaviour, and expectations of respondents of the survey have come up specifically with respect to energy and water saving and visible pollutions (municipal solid wastes). The figures (Figs. 5.1, 5.2, and 5.3) present the expectation of the respondents towards government's action for reducing carbon emission across zones, viz. the rural, peri-urban, and urban areas of 13 locations of the country. Figure 6.1 describes the household's expectation from government to introduce measures to reduce carbon emission. In all the zones rural, peri-urban, and urban, more than 60% of the households expect initiation of awareness programmes regarding emissions and its impacts followed by stringent rule imposition by the government.

Fig. 5.1 Measures expected from government (Rural)

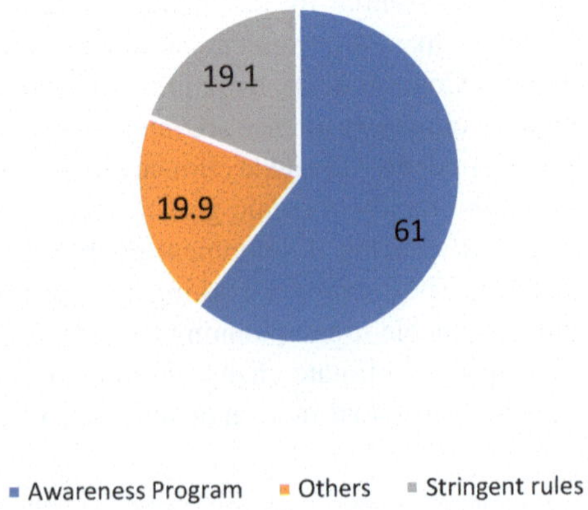

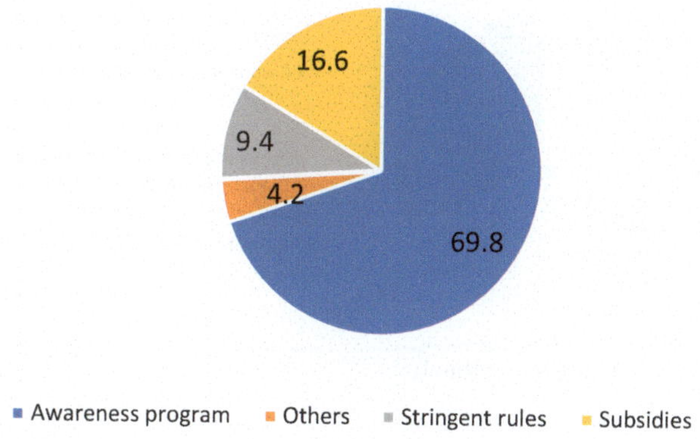

Fig. 5.2 Measures expected from government (Peri-urban)

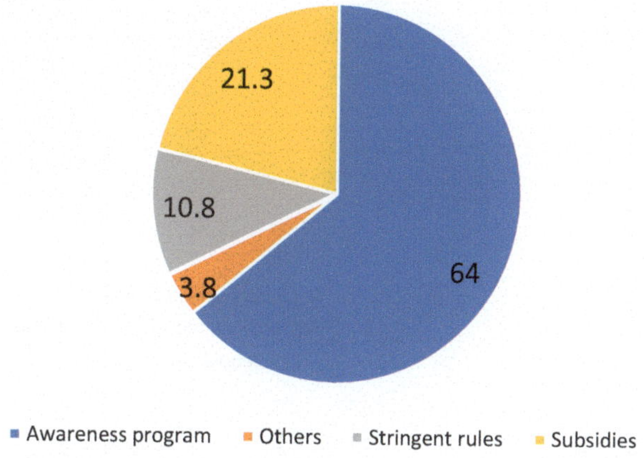

Fig. 5.3 Measures expected from government (Urban)

5.2 Environmental Awareness

With respect to awareness about various terminologies related to climate change and GHG emission, zones (urban, rural, and peri-urban) have exhibited an impact on people's knowledge (Figs. 5.4, 5.5, and 5.6). The figures describe that in urban zone, 51.4% of the respondents are aware of the carbon emission reduction, whereas in peri-urban, and rural zones 52.2%, and 67.6% of the respondents are aware of the ecological footprint/household carbon footprint, respectively, followed by household energy saving with 35.8%, and 40.8% in rural and peri-urban zones.

While the respondents were asked about the actions, they take up consciously in their day-today life which can be counted as carbon-friendly behaviour, saving electricity, and water emerged as the top initiative undertaken by almost 34% households. More than 90% of the households across all the locations reported practising one or more acts (Fig. 5.7) to reduce environmental impact due to consumption at

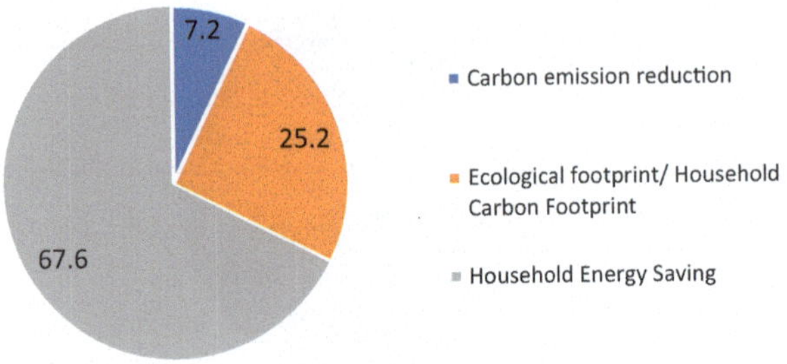

Fig. 5.4 Environmental awareness (Rural)

Fig. 5.5 Environmental awareness (Peri-urban)

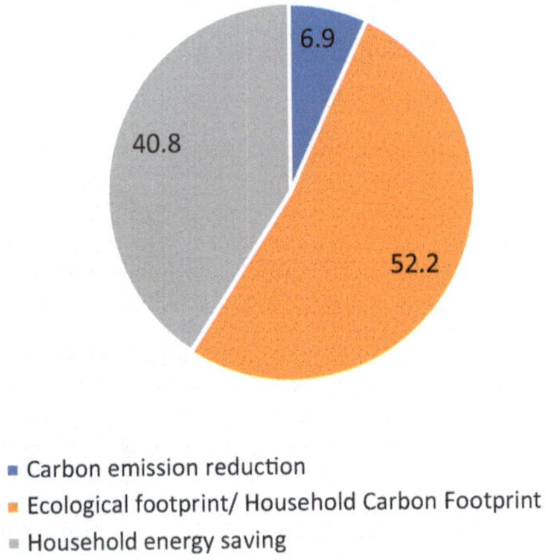

Fig. 5.6 Environmental awareness (Urban)

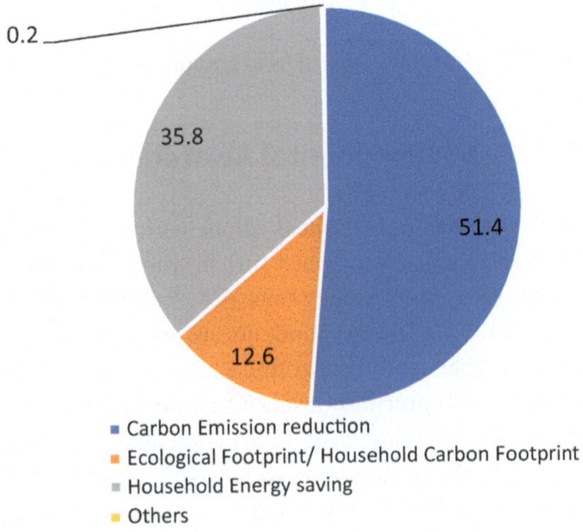

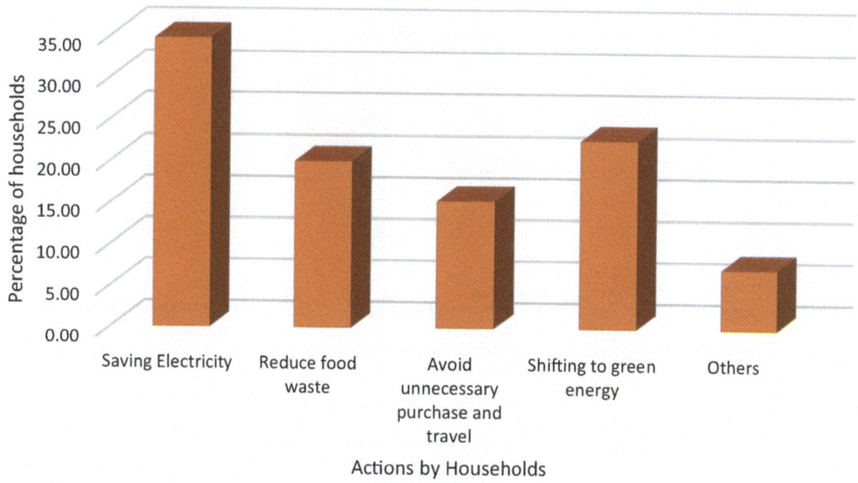

Fig. 5.7 Household's action to reduce carbon footprint

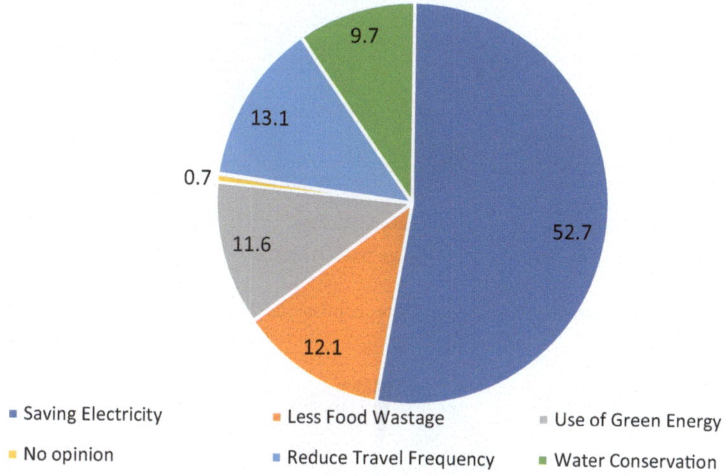

Fig. 5.8 Measures considered to reduce carbon emissions (Rural)

household level. Figure 5.3 describes measures that respondents can consider to take up to reduce carbon emission. In rural zone, about 27.7% of respondents are ready to take up conservation of water as a measure to reduce carbon emission whereas 52.7%, and 55.1% of respondents opted for saving electricity in peri-urban and urban zones, respectively.

Graphs describe measures (Figs. 5.8, 5.9, and 5.10) and barriers (Figs. 5.11, 5.12, and 5.13) to adopt carbon emission reduction measures. In rural area, about 25.8% of the respondents pointed out "cost" as a barrier to adopt the measures to reduce carbon emission. In rural India, as the emission reduction has not been

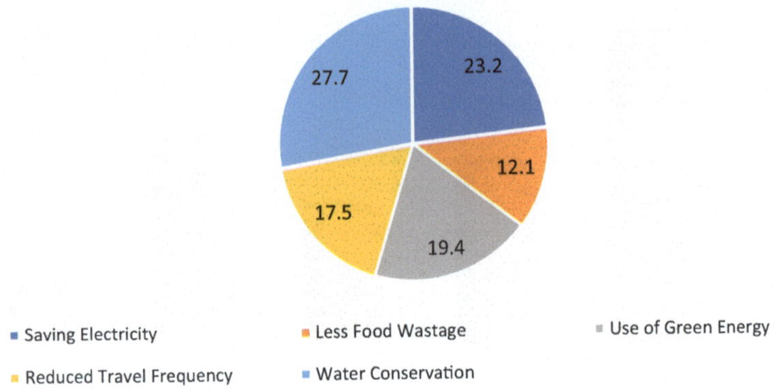

Fig. 5.9 Measures considered to reduce carbon emissions (Peri-urban)

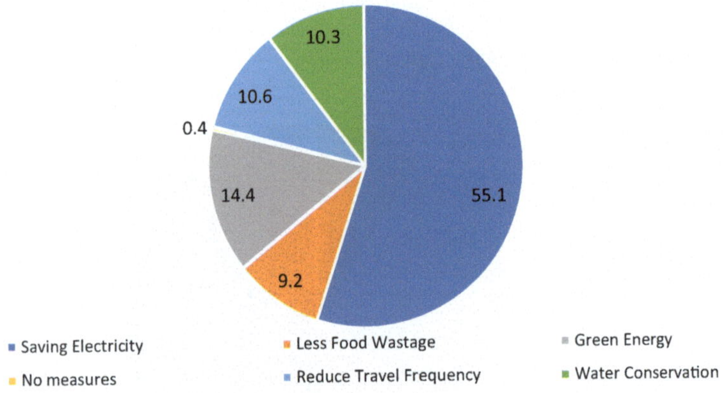

Fig. 5.10 Measures considered to reduce carbon emissions (Urban)

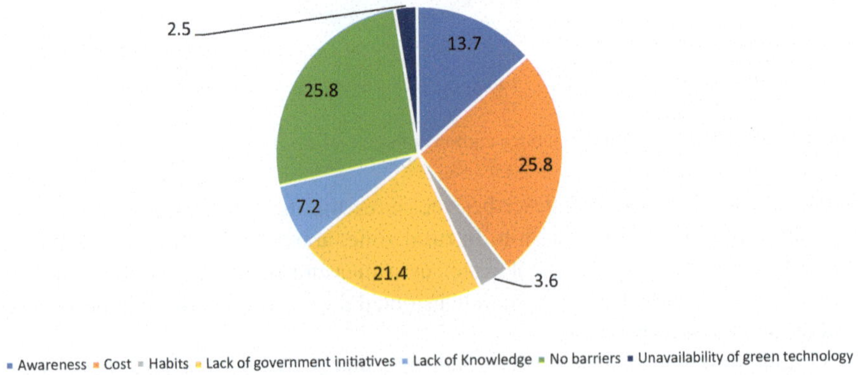

Fig. 5.11 Barriers in adopting measures (Rural)

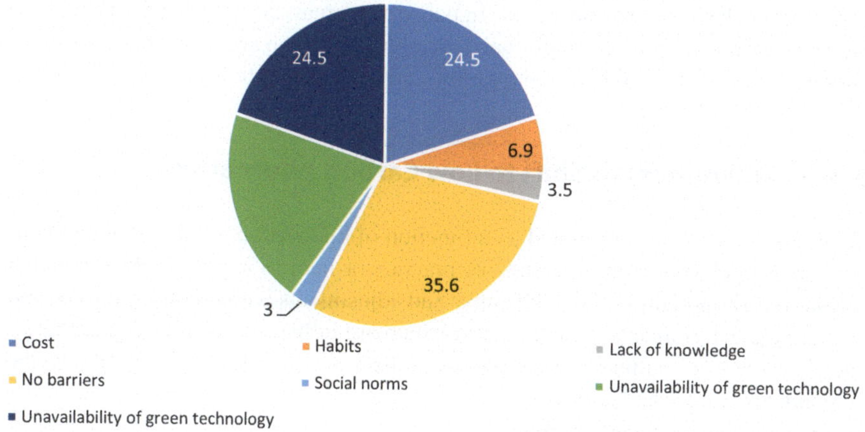

Fig. 5.12 Barriers in adopting measures (Peri-urban)

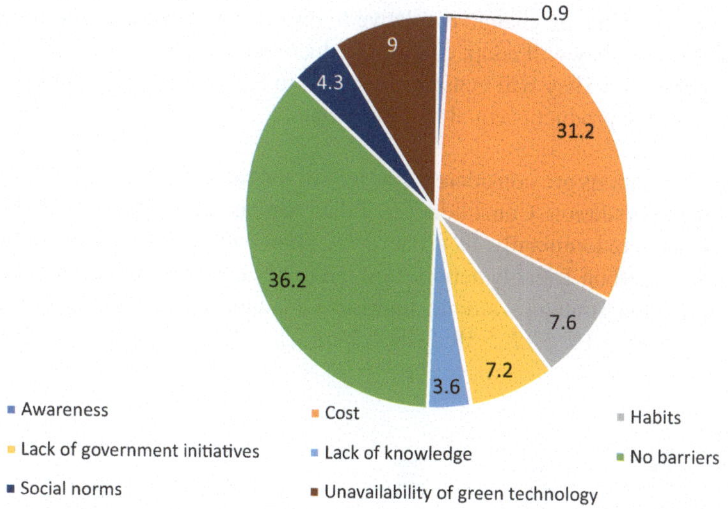

Fig. 5.13 Barriers in adopting measures (Urban)

emerged as a concern among middle- and low-income groups, rather has a potential increase in carbon footprint, with special respect to food HCF for achieving sustainable development goals. Thus, a confused response in rural areas was received, mixing the barriers in economic well-being and betterment of the environment, which produces a larger share of respondents attributing government inefficiency in overall rural development. On the other hand, significant share of the respondents, 35.6% and 36.2%, do not recognize any barriers to adopt the measures to reduce carbon emission in peri-urban and urban areas, respectively. In rural areas, almost all the options listed as barriers to adoption of low carbon lifestyles and household consumption have been as prevalent. In peri-urban areas, costs and lack of access to greener alternatives are two major barriers reported. Urban areas also have a similar pattern of obstacles in transitioning to low carbon consumption.

5.3 Willingness to Shift to Low Carbon Alternatives

Willingness has been formulated as a function of knowledge, motivation, and consideration to change from one state or practice or item to another with somewhat similar utility and purpose. Compromise and adjustments with quality, comfort, and cost are the determinants of willingness which are influenced by a large number of social, economic, cultural, political, and religious attributes. In the survey, the respondents were asked about

The respondents mostly believe:

- Knowledge—The citizens believe that environmental impacts are frequently overstated.
- Motivation—Climate change is relative to disease and health, so for betterment of their health they will adopt emission reduction methods.
- Consideration—They will consider to switch to low emission alternative even if the cost of the product is equal to current one.

Policy regulations are considered as effective instrument to reduce emissions by the surveyed population. Consolidation of data who are above average HCF of our estimates and predominantly in urban areas shows imposition of restrictions or facilitation of carbon-friendly actions through policy can have immense impacts on carbon reduction strategies among high- and medium-income urban households (Figs. 5.14, 5.15, 5.16, 5.17, 5.18, 5.19, 5.20, 5.21, 5.22, 5.23, and 5.24). While the

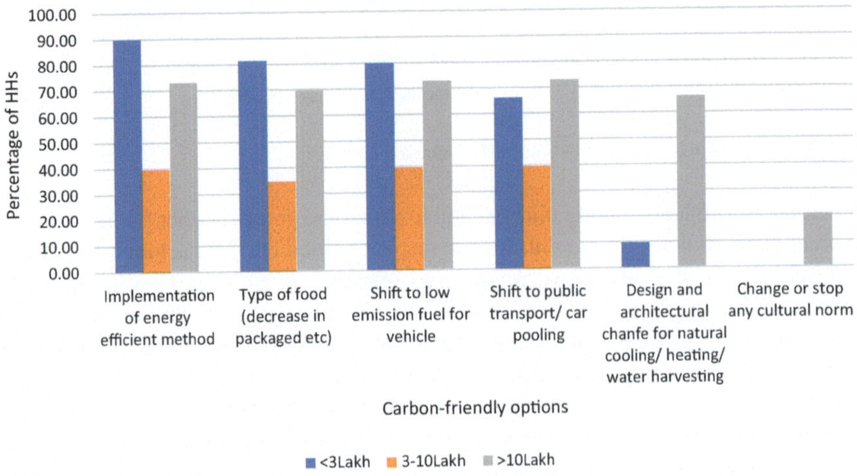

Fig. 5.14 Household willingness to shift to low-carbon alternatives only if there is strict imposition by government

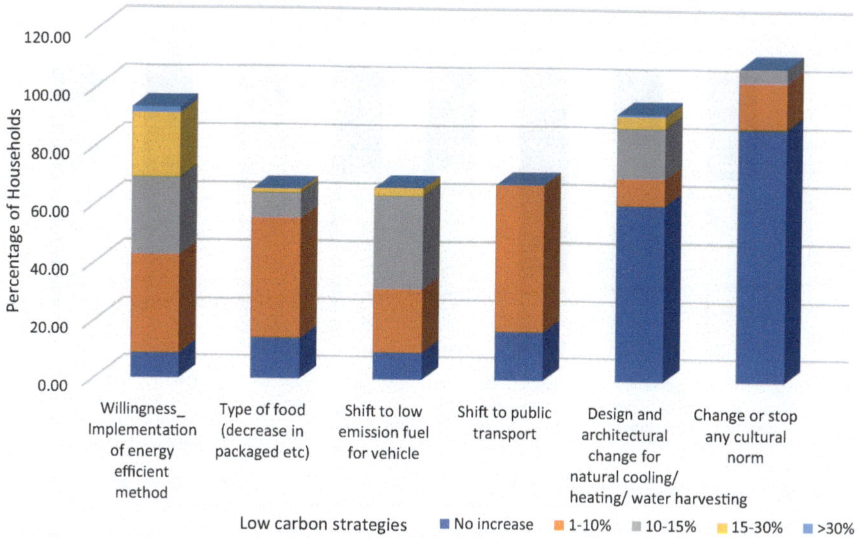

Fig. 5.15 Respondent's willingness to increase expenditure or to compromise with comfort

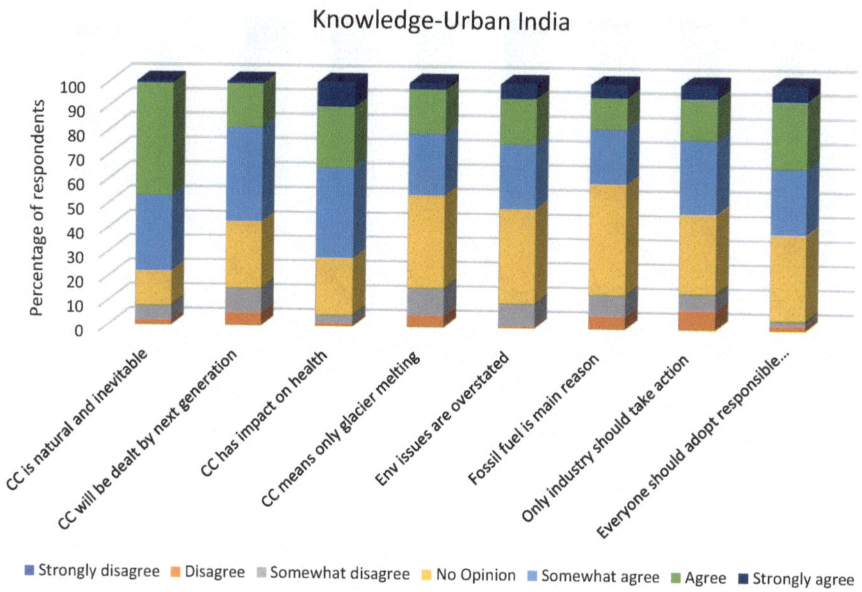

Fig. 5.16 Opinion of the urban population (knowledge)

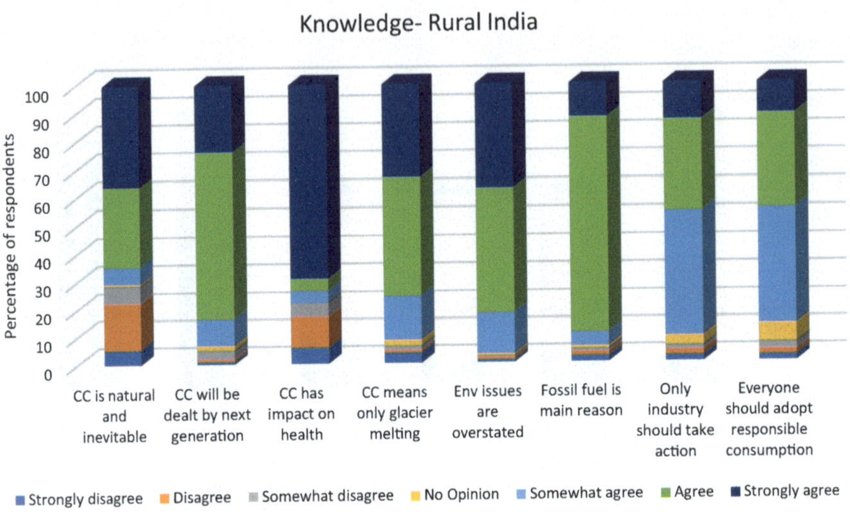

Fig. 5.17 Opinion of the rural population (knowledge)

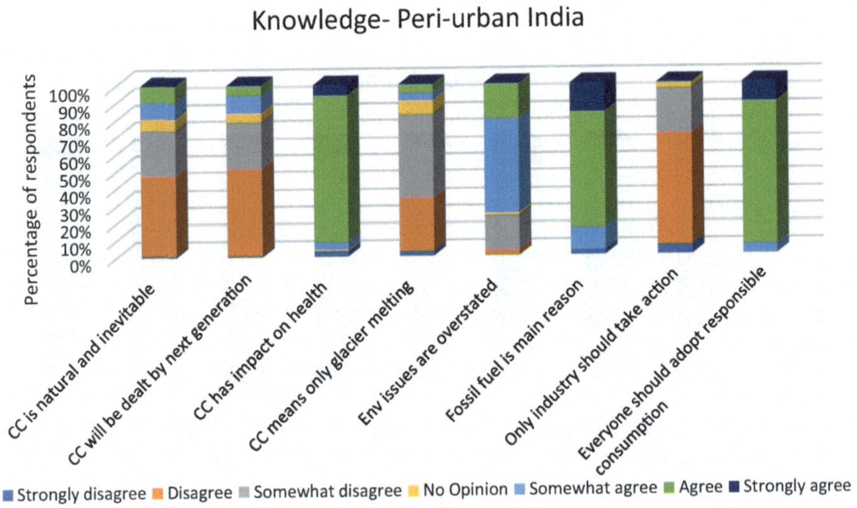

Fig. 5.18 Opinion of the peri-urban population (knowledge)

urban economically well-to-do households were given a few options and asked about their willingness to increase expenditure or compromise with comfort/norms to achieve low-carbon lifestyle, responses were received about more flexibility in terms of spending money and comfort for better green energy source and public transport if conveniently available.

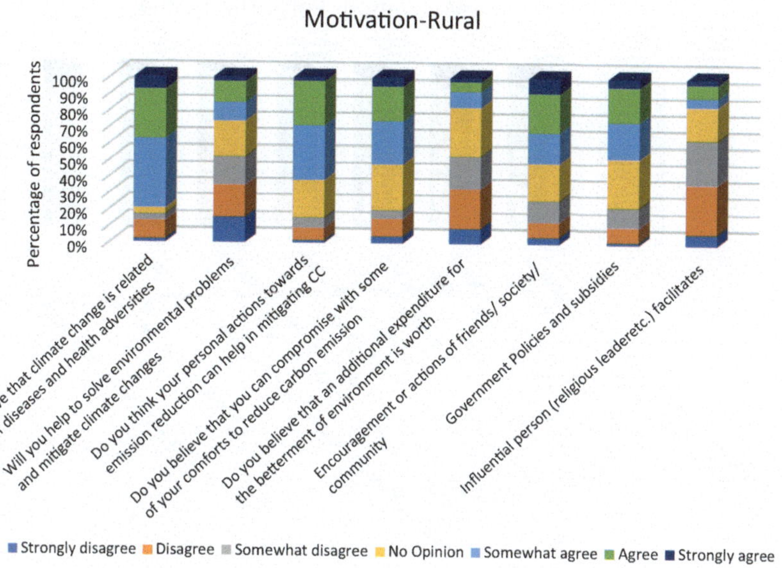

Fig. 5.19 Opinion of the rural population (motivation)

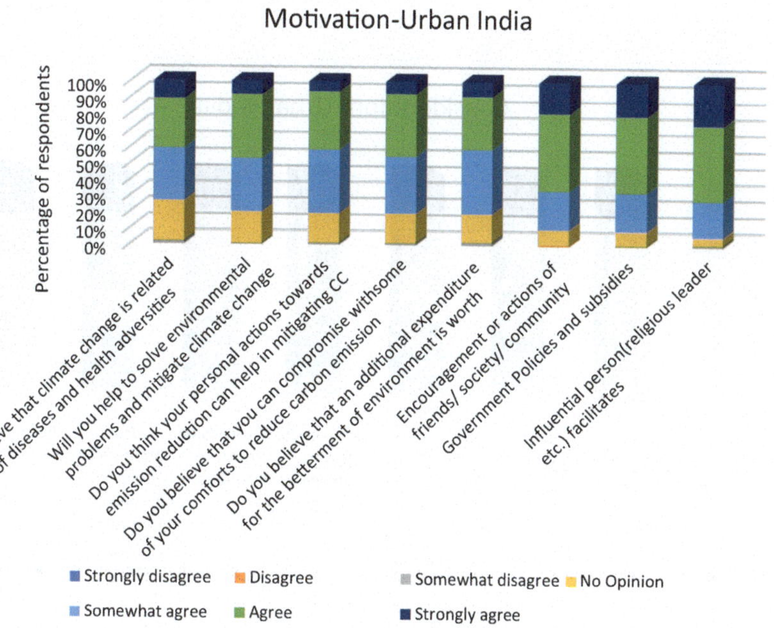

Fig. 5.20 Opinion of the urban population (motivation)

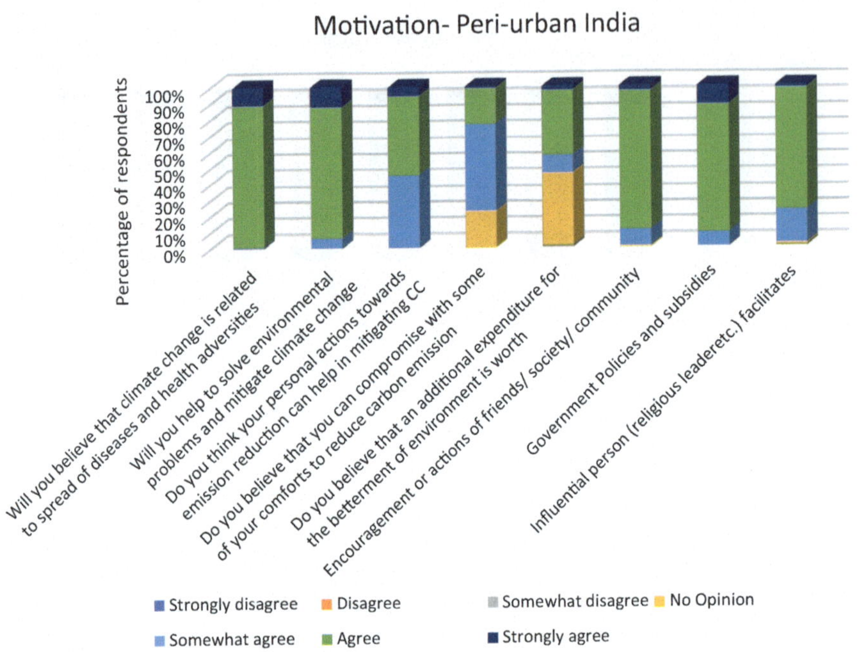

Fig. 5.21 Opinion of the peri-urban population (motivation)

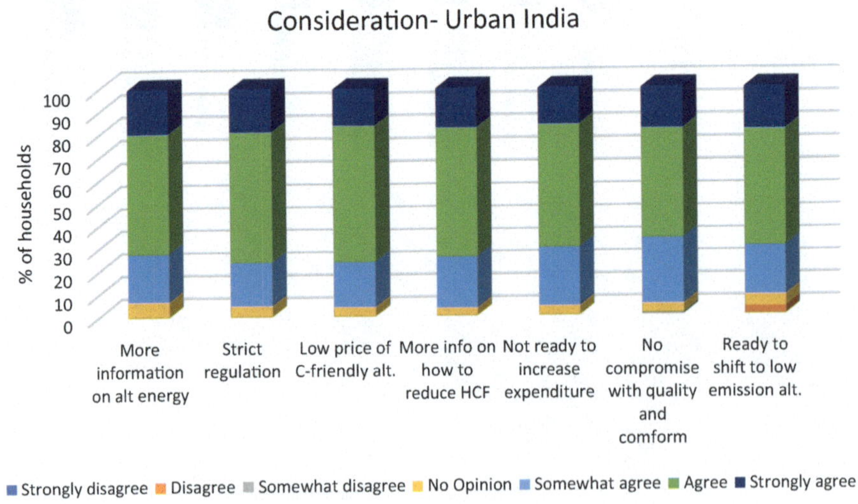

Fig. 5.22 Opinion of the urban population (consideration)

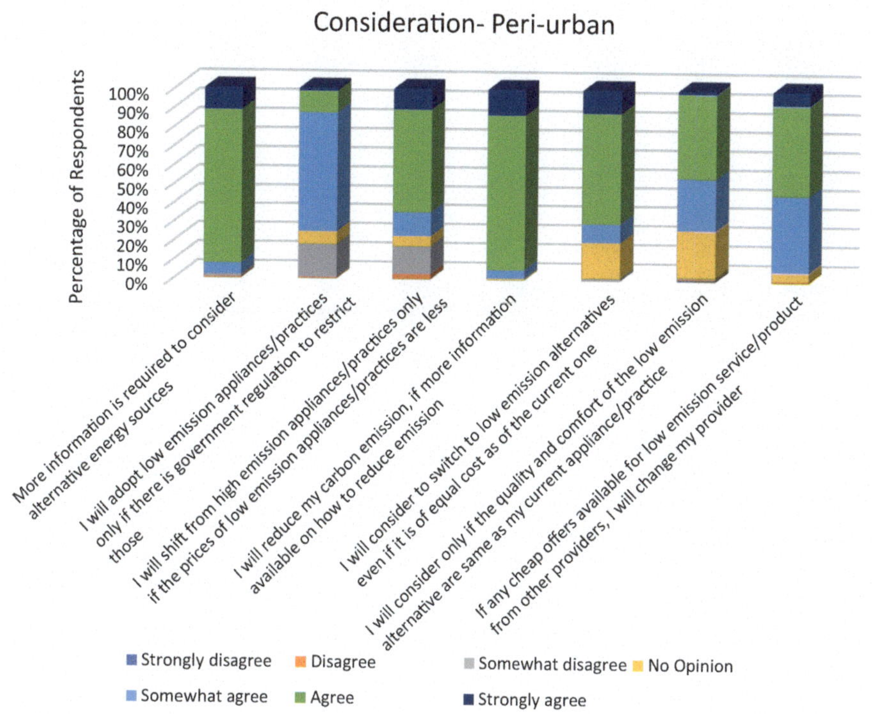

Fig. 5.23 Opinion of the peri-urban population (consideration)

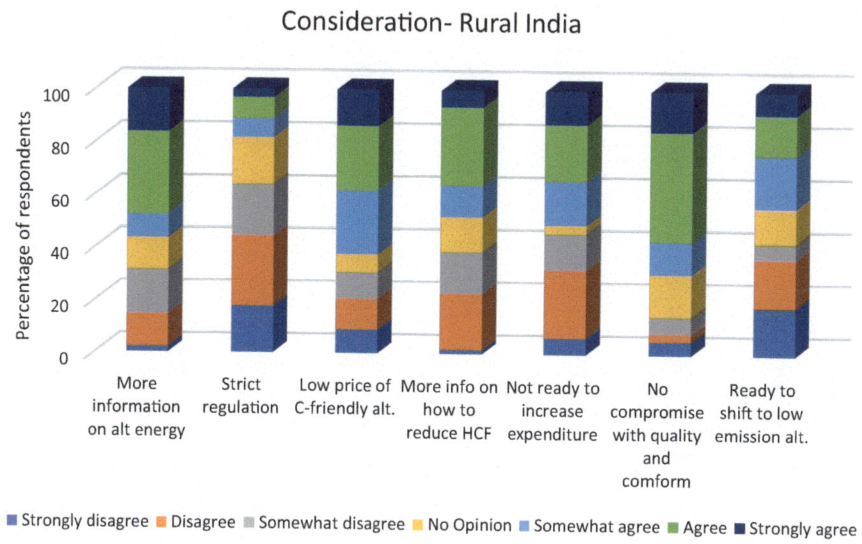

Fig. 5.24 Opinion of the rural population (consideration)

Influence of Affluence and Other Determinants of HCF

6

The household carbon footprint (HCF) is a measure of the total greenhouse gas emissions generated directly and indirectly by households. Various factors influence the HCF, with affluence being a significant determinant. Understanding how affluence and other determinants affect HCF is crucial for developing targeted strategies to reduce emissions and promote sustainable living. Affluent households tend to consume more goods and services, leading to higher carbon emissions. Greater disposable income often results in the purchase of larger houses, more electronic devices, and luxury items, all of which contribute to higher energy use and emissions. The adoption of modern conveniences and technologies, such as smart home systems and home entertainment centres, also contributes to higher energy consumption. Dietary choices influenced by affluence can lead to higher emissions. Affluent households tend to consume more meat and dairy products, which have a higher carbon footprint compared to plant-based foods. The preference for imported and out-of-season foods among wealthier households also adds to their carbon footprint due to the transportation emissions involved in bringing these products to market. Governments and organizations can support this by providing incentives for sustainable products and implementing policies that promote responsible consumption. The other determinants of HCF include geographic location, household size and composition, education and awareness, and access to technology and infrastructure. By understanding these determinants, targeted strategies can be developed to promote sustainable consumption, enhance energy efficiency, increase renewable energy use, and improve public transportation and infrastructure. Through collective efforts and informed policy-making, it is possible to reduce household carbon footprints and contribute to global climate mitigation efforts.

S. Nautiyal et al., *Determinants and Stratification of Household Carbon Footprint*,
https://doi.org/10.1007/978-981-96-2739-4_6

6.1 Household Expenditure and HCF

Table 6.1 describes the regression results of all India expenditure with HCF. Simple linear regression was performed between expenditure categories and total household carbon footprint. It can be noted that LPG expenditure is statistically significant at 5% level. From the result coefficients explain that hundred rupees increase in LPG expenditure leads to decrease in −0.01 kg per CO_2e of total HCF. On the other hand, hundred rupees increase in household total expenditure leads to increase in 0.8 kg per CO_2e of HCF, hundred rupees increase in household food expenditure leads to increase in 0.1 kg per CO_2e of HCF, hundred rupees increase in travel expenditure leads to increase in 0.3 kg per CO_2e of HCF, and hundred rupees increase in electricity expenditure leads to increase in 0.2 kg per CO_2e of HCF.

The given figure (Fig. 6.1) shows a positive relationship between the annual expenditure and the total household's emissions, which can be observed in the fit line (the terms figure and graph have been used interchangeably). The strength of the relationship between the two variables is weak as the values are scattered. The predictive emissions suggest a uniform increase in emissions with an increase in expenditure.

Figure 6.2 shows more uniform concentration between the electricity emissions and the household expenditure than the previous figure. In this figure, the steep slope suggests a positive correlation between the variables and the prediction emissions shown increase in electricity.

The relationship between food and expenditure shows a weaker relationship with gradual inclination (Fig. 6.3). Isolated points can also be observed in the scatterplot. Therefore, expenses on food do not necessarily signify an increment in emissions as the consumption of food is limited to members of a household.

Figure 6.4 indicates a steep rise in the emissions with an increase in transport and travel. The relationship between the two variables deems to be positive. Therefore, each trip a household makes leads to *an* increase in the total emissions. The predicted values suggest further increase in the expenses will lead to higher emissions.

Table 6.2 is the description of the variables which provides an outline of independent variables influencing dependent variables for the given study areas. The "ticks" shows significant relationship between the two variable types for the two areas. Emissions from energy use, food, travel, and waste of the households of Joshimath region is predicated upon different independent variables but mostly by household

Table 6.1 Increase in carbon footprint (CF) with per rupee increase in total household expenditure

Dependent variables	Coefficients	Standard error	*P* Value	*R* Squared
Household total expenditure	0.008	0.001	6.93	0.0919
Food expenditure	0.001	0.0003	1.76	0.0999
Travel expenditure	0.003	0.0006	3.80	0.0954
Electricity expenditure	0.002	0.0002	7.19	0.2180
LPG expenditure	−0.0001	0.00006	0.01**	0.5186

**5% level of significance

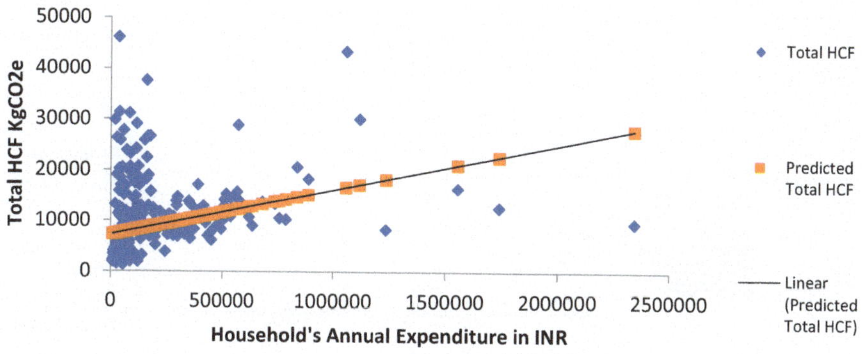

Fig. 6.1 Total expenditure vs total HCF

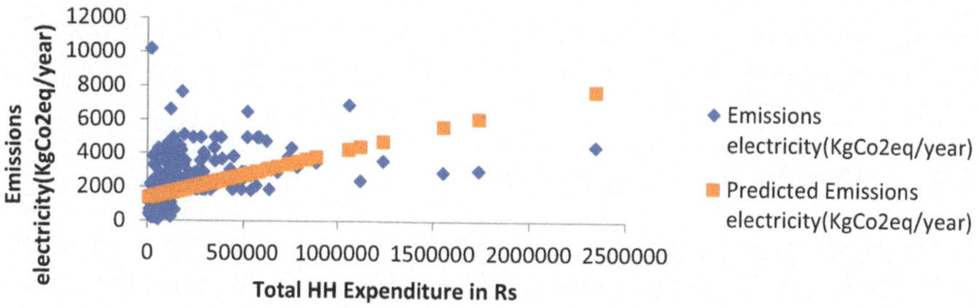

Fig. 6.2 Total expenditure vs electricity

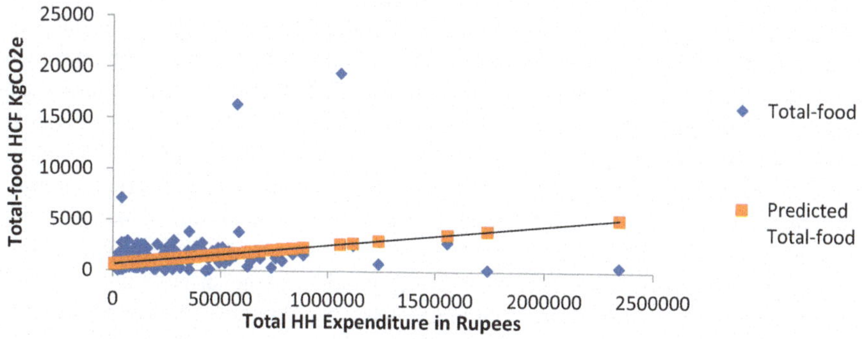

Fig. 6.3 Total expenditure vs food

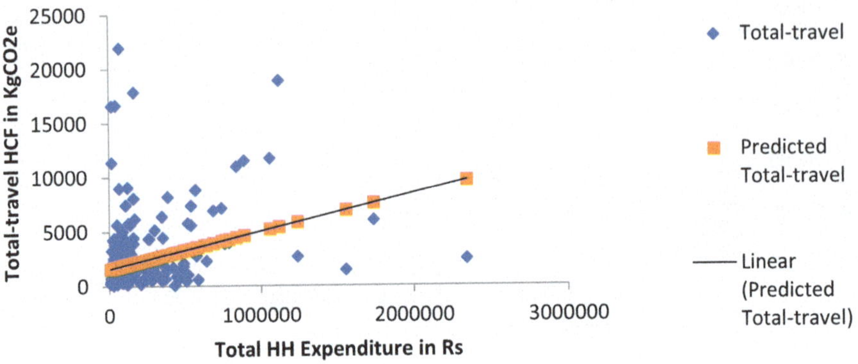

Fig. 6.4 Total expenditure vs travel

income. Whereas for the Mangalore region households, again, income is responsible for the emissions from energy use, waste, and internet use.

6.2 All India Total Regression

The Table 6.3 provides the regression analysis between multiple dependent and independent variables and their effect on total household emissions for all the combined locations. Figures in the table represent coefficients and figures in the parentheses represent standard errors. Coefficients denote that one unit increase in particular variable leads to change in dependent variable. For all the dependent variables in the analysis, we can observe the changes in the positive coefficients and significance level at 1% ($p < 0.01$). The changes which follow under the dependent variable HCF are rural, family size, income groups 10 lakhs to 30 lakhs and above 30 lakhs. For the dependent variable (other non-durables), the significant independent variables are urban, income groups 5–10 lakhs, 10–30 lakhs, and above 30 lakhs. For the dependent variable (waste), the significant independent variables are family size and income group 10–30 lakhs. Dependent variable used in the regression is petrol. It has been updated for other variables as well. For the dependent variable electricity, the significant independent variables are rural, urban, family size, income groups 1–3 lakhs, 5–10 lakhs, 10–30 lakhs, and above 30 lakhs. For the dependent variable (LPG), the significant independent variable is family size. For the dependent variable (biomass), the significant independent variables are rural and family size. For the dependent variable (cereals), the significant independent variable is family size. For the dependent variable (milk and milk products), the significant independent variables are rural, family size, income groups 5–10 lakhs and 10–30 lakhs. For the dependent variable (vegetables and fruits), the significant variable is family size. For the dependent variables (fish, meat, internet, public travel

Table 6.2 Factors having statistically significant impact on emissions at 95% confidence level

Independent variables	Energy		Food		Travel		Biodegradable waste		Non-biodegradable		Internet use	
	Joshimath	Mangalore	Joshimath	Mangalore	Joshimath	Mangalore	Joshimath	Mangalore	Joshimath	Mangalore	Joshimath	Mangalore
Family size	✓		✓	✓	✓	✓						✓
Education	✓		✓									
HH income	✓	✓	✓				✓	✓	✓	✓		✓
landholding size			✓			✓	✓	✓	✓	✓		
Built-up space		✓					✓	✓				
Occupation				✓		✓			✓	✓		
Housing type		✓		✓					✓	✓		

Table 6.3 All India regression for all the variables

Dependent variables	Rural	Urban	Family size	50 k to 1L	1–3 L	3–5 L	5–10 L	10–30 L	Above 30 L	R^2
HCF	801.05*** (263.42)	439.64 (289.14)	603.16*** (57.55)	−1026.41*** (337.92)	−256.08 (347.90)	76.50 (397.10)	1163.53** (467.55)	5893.78*** (625.76)	11980.38*** (2115.58)	0.1462
Other non-durables	16.67 (14.35)	41.09*** (15.75)	5.38* (3.13)	4.28 (18.41)	42.68** (18.96)	24.85 (21.64)	105.19*** (25.48)	798.96*** (34.10)	1386.19*** (115.30)	0.3393
Waste-total	129.37** (57.35)	95.31 (62.95)	122.77*** (12.53)	−158.69** (73.57)	−116.07 (75.75)	−95.27 (86.46)	−215.21** (101.80)	424.04*** (136.25)	30.52 (460.64)	0.0671
Petrol	−56.94 (127.69)	19.65 (140.15)	62.77** (27.89)	68.78 (163.79)	428.97** (168.63)	768.11*** (192.48)	594.35*** (226.63)	1170.07*** (303.32)	1657.26 (1025.47)	0.0285
Electricity	748.82*** (94.23)	306.73*** (103.43)	168.04*** (20.58)	243.38** (120.88)	341.38*** (124.45)	189.50 (142.05)	588.64*** (167.25)	2135.61*** (223.84)	2130.09*** (756.78)	0.1166
LPG	−281.65*** (104.40)	−22.00 (114.59)	123.58*** (22.81)	−467.24*** (133.93)	−411.96*** (137.88)	−309.62** (157.38)	−138.06 (185.30)	−662.16*** (248.01)	−717.36 (838.48)	0.0298
Biomass	339.64*** (57.22)	−92.84 (62.80)	45.02*** (12.50)	−760.32*** (73.40)	−951.62*** (75.57)	−920.40*** (86.26)	−996.34*** (101.56)	−939.45*** (135.93)	−908.09** (459.56)	0.1661
Cereals	4.44 (3.90)	−9.85** (4.28)	12.38*** (0.85)	−17.16*** (5.00)	−26.12*** (5.15)	−29.24*** (5.88)	−30.18*** (6.92)	−36.76*** (9.26)	−31.19 (31.32)	0.1452
Pulses	−17.51 (11.92)	−14.85 (13.08)	1.92 (2.60)	12.97 (15.29)	−1.26 (15.74)	−6.49 (17.97)	−13.85 (21.16)	17.43 (28.32)	19.77 (95.74)	0.0033
Milk and milk products	76.95*** (14.78)	14.81 (16.23)	14.17*** (3.23)	−5.93 (18.96)	45.15** (19.53)	40.51* (22.29)	77.09*** (26.24)	353.95*** (35.12)	199.93* (118.76)	0.0839

Veg+ fruits	4.91** (2.02)	−3.35 (2.22)	2.59*** (0.44)	−14.66*** (2.59)	−10.96*** (2.67)	−10.78*** (3.05)	−13.63*** (3.59)	1.47 (4.81)	−0.95 (16.27)	0.0512
Chicken	67.10 (58.46)	7.89 (64.16)	10.58 (12.77)	51.84 (74.99)	15.73 (77.20)	114.31 (88.12)	36.97 (103.76)	83.26 (138.87)	93.11 (469.50)	0.0026
Fish	−18.60 (27.39)	6.06 (30.07)	1.97 (5.98)	−2.33 (35.14)	0.01 (36.18)	−2.21 (41.29)	2.59 (48.62)	306.38*** (65.08)	56.36 (220.02)	0.0191
Meat	−68.62* (38.23)	55.06 (41.96)	15.21* (8.35)	−14.36 (49.04)	14.22 (50.49)	1.33 (57.63)	128.34* (67.86)	371.12*** (90.82)	753.46** (307.06)	0.0347
Internet	−11.12*** (2.18)	−0.94 (2.39)	5.33*** (0.47)	−8.39*** (2.80)	−6.36** (2.88)	−7.01 (3.29)	9.04** (3.87)	65.42*** (5.18)	68.45*** (17.53)	0.22
Public travel	−120.85*** (40.10)	−34.54 (44.01)	−4.07 (8.76)	−0.24 (51.44)	72.08 (52.96)	22.56 (60.45)	285.76*** (71.17)	146.01 (95.26)	3129.01*** (322.06)	0.0745
Private travel	−28.23 (56.85)	30.33 (62.40)	10.06 (12.42)	37.37 (72.93)	265.34** (75.08)	271.49*** (85.70)	637.63*** (100.90)	859.41*** (135.05)	2727.60*** (456.59)	0.0746

*$p < 0.1$
**$p < 0.05$
***$p < 0.01$

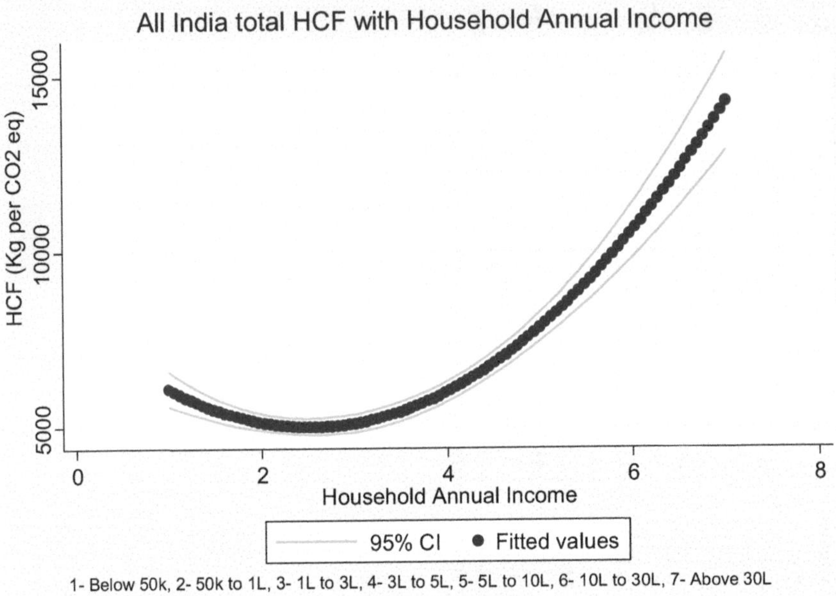

Fig. 6.5 Scatter plot of income with HCF

and private travel), the significant independent variables are family size, income groups 3–5 lakhs, 5–10 lakhs, 10–30 lakhs, and above 30 lakhs. Broadly, it can be inferred that the larger the family more the emissions lead to increase in per capita emissions. Also, the higher income groups are the major emitters when compared to the lower income groups.

The given scatter plot (Fig. 6.5) depicts the relationship between the variables Household Carbon Footprint (HCF) which is at the *y*-axis and the Annual Household Income at the *x*-axis. The points in the figure suggest clear relationship between the emissions and income. A positive correlation can be observed in the plot with upward slope, variable on the *y*-axis tends to increase when there is an increase in the variable on *x*-axis.

The given scatter plot (Fig. 6.6) depicts the relationship between the two variables Household Carbon Footprint which is at the *y*-axis and family size which is at the *x*-axis. The points in the graph suggest clear relationship between the emissions and the family size. A positive correlation can be observed in the plot with upward movement of the slope, variable on the *y*-axis tends to increase when there is an increase in the variable on *x*-axis.

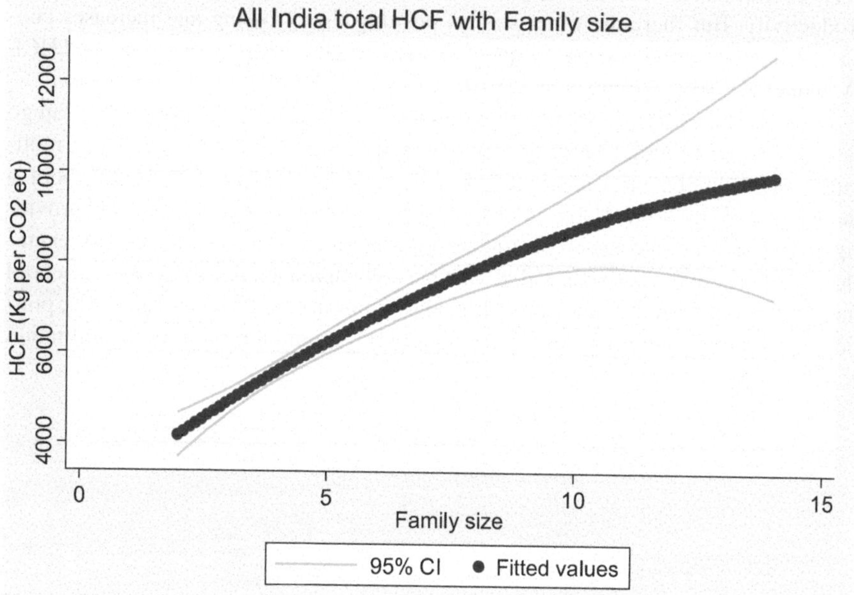

Fig. 6.6 Scatter plot of family size with HCF

6.3 Discussion

72% of global greenhouse gas emissions and 70% of global CO_2 emissions are directly held accountable by household consumption activities. This signifies the important role played by household participation (Baiocchi et al. 2010; Hertwich and Peters 2009; Niamir et al. 2020).

Changing household behavioural patterns can reduce global carbon emissions connected to energy use and transition to low-carbon products and services. This very reason highlights the significant role of household consumption in global carbon emissions (Alfredsson 2004).

A major factor influencing HCF levels, in addition to income, can be considered the number of household members. Previous research shows that an indirect hike in HCF levels can be noted when household size increases, which corresponds to great product consumption (Irfany and Klasen 2017; Levay et al. 2021). The working members and number of children are directly related to HCE levels; hence, the composition of household members plays a crucial role in HCF. An increase in energy and food consumption can be assured due to the presence of children in household activities. At the same time, household income and the capacity for consumption are directly proportional to the number of working members (Koide et al. 2019; Levay et al. 2021). Even though an increase in household head production can lead to a rise in HCF, it declines after attaining the optimal point, coinciding with a decrease in productivity. But there are other studies positing that growing age increases consumption needs and that there is a positive linear relationship between age and HCF (Adnan et al. 2018; Hirano et al. 2016).

A low-carbon lifestyle should be promised by a shift in social norms that can go beyond the behavioural changes accompanied by economic benefits. To boost motivation individually, economic incentives can be implemented to bring about a change in their lifestyle. Using mini solar systems for power generation and providing hydrogen for cooking and heating necessities can be considered as major examples of alternative approaches that can be considered to decarbonize household energy use. In addition to the innovation and promotion of these technologies, policies, investments, and financial support should be granted for the smooth facilitation of these approaches.

Impact of Changing Landscape Dynamics Along Rural–Urban Interface 7

The interface between rural and urban areas, often referred to as the rural–urban fringe, is a dynamic and evolving landscape. This transition zone is characterized by a blend of rural and urban features, where the impact of urbanization is particularly pronounced. As urban areas expand, the dynamics of the rural–urban interface change, leading to significant environmental, social, and economic consequences. The changing dynamics along rural–urban interfaces have profound impacts on the environment, society, and economy. Urban expansion leads to the loss of agricultural land, biodiversity decline, and water resource challenges, while also bringing improved infrastructure and services to rural areas. Social impacts include community displacement and cultural changes, as well as increased housing demand and urban sprawl. Economically, the rural–urban interface presents both challenges and opportunities, affecting land values, agricultural productivity, and economic diversification.

7.1 Spatio-temporal Changes

Land use land cover data is the basis of global-change research and regional governmental management. The changes in land surface directly affect the environment. Rapid development of infrastructure and economic conditions over the past few decades has culminated into the ability of humans to change the land cover significantly.

From the land use land cover classification results in Tirupattur district (Table 7.1, Figs. 7.1, and 7.2), it was observed that the built-up area has significantly increased from 2000 to 2020. Built-up area has increased by 10.44% in urban zone, 1.24% in peri-urban, and 1.02% in rural zone. The capacity of water bodies has also increased by 0.54% in urban zone, 0.51% in peri-urban, and 0.04% in rural zone. Vegetation has been affected in all regions, vegetation in urban zone has decreased by 2.6%, in peri-urban by 0.04%, and in rural zone by 9.93%. Fallow land has also decreased and cropland is also affected, mostly in urban areas, it decreased by 15.59% in urban zone, 13.34% in peri-urban, and 8.49% in rural zone. Decrease in cropland and vegetation may directly impact the environment, but a positive outcome can be observed in the form of increased forest land, mainly in rural zone.

The land use land cover classification of Bhopal (Table 7.2, Figs. 7.3 and 7.4) results show that the built-up area has increased by 9.24% in urban zone, whereas it has increased by 7.83% in rural zone. The results of Bhopal city clearly show that

S. Nautiyal et al., *Determinants and Stratification of Household Carbon Footprint*,
https://doi.org/10.1007/978-981-96-2739-4_7

Table 7.1 Spatio-temporal change in percentage of area under land use land cover classes (Tirupattur)

Tirupattur	2000			2020		
Features	Urban	Peri-urban	Rural	Urban	Peri-urban	Rural
Water bodies	0.30	0.07	0.05	0.84	0.58	0.09
Forest	0.12	4.19	7.34	1.39	4.42	9.76
Built-up	4.22	1.32	1.94	14.66	2.56	2.96
Rocky area	3.16	2.51	4.12	3.76	4.64	4.05
Cropland	27.19	38.96	32.73	11.60	25.62	24.24
Fallow land	10.39	7.19	9.41	8.17	8.19	8.38
Vegetation	24.04	13.93	20.62	21.44	13.89	11.51
Other features	30.58	31.84	23.79	37.91	40.09	38.98

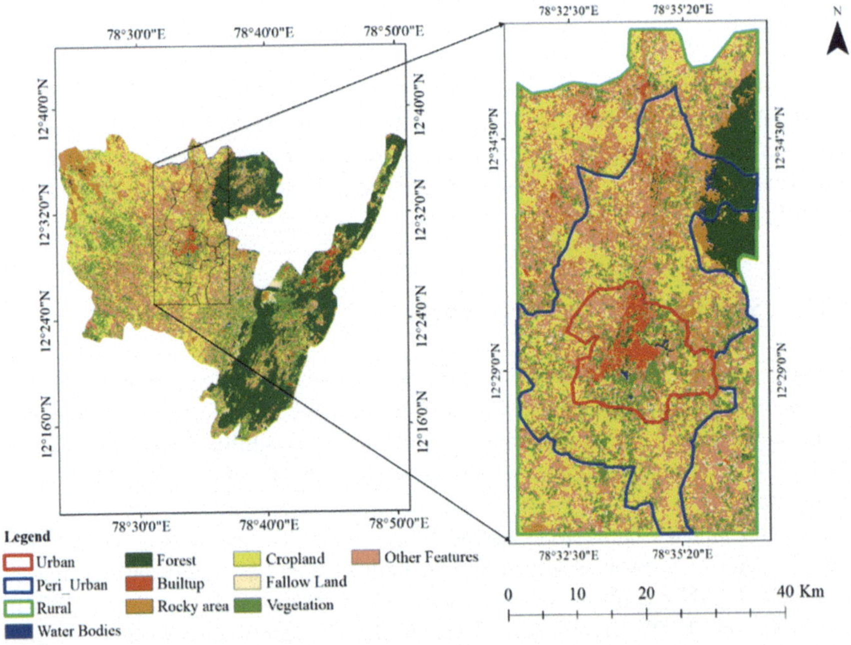

Fig. 7.1 LULC of Tirupattur (2000)

the urbanization has increased rapidly between the period of 2000 and 2020. The Cropland is reduced by 2.8% in rural zone and 2.74% in urban zone. The fallow land has also decreased and vegetation is also affected. The changes in fallow land and vegetation are mainly due to the increase in built-up areas.

Noida is one of the planned cities in India and holds the tag of greenest city. The land use/land cover statistics (Table 7.3, Figs. 7.5 and 7.6) shows that the built-up area has increased by 17.38% in urban zone of Noida, whereas 18.23% area has increased in rural zone. Vegetation and fallow land have been affected over a period of time. Cropland is also affected, mainly in rural zone. Cropland has decreased by

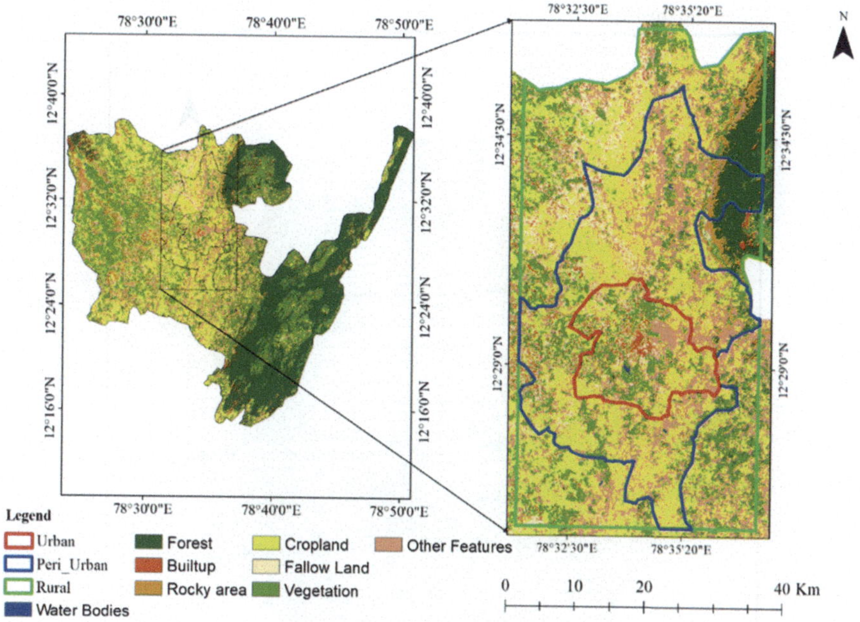

Fig. 7.2 LULC of Tirupattur (2020)

Table 7.2 Spatio-temporal change in percentage of area under land use land cover classes (Bhopal)

Bhopal	2000		2020	
Features	Urban	Rural	Urban	Rural
Water bodies	5.13	0.54	6.43	0.03
Built-up	45.42	2.32	54.67	10.13
Cropland	14.96	52.76	12.23	49.95
Fallow land	30.78	41.83	24.69	38.25
Vegetation	3.71	2.55	1.99	1.64

33.06% in rural zone and 20.98% in urban zone. The decreased cropland may affect the food security of the region.

The findings of the land use/land cover results in Ashta region (Table 7.4, Figs. 7.7 and 7.8) shows the water bodies getting affected in current scenario. Built-up area has increased by 4.15% between the year 2000 and 2020. Fallow land and vegetation are contracting with the expansion of developmental activities, but positive results in cropland have been analysed, which shows 11.79% increment.

Joshimath is one of the famous cities of Chamoli district in Uttarakhand. The findings of land use/land cover show (Table 7.5) that the dense forest is getting affected from 2000 to 2020 by 20.1% and built-up area has increased by 3.32%. Vegetation is also diminishing with time and has decreased by 2.28%. Perforated forest has also decreased by 0.84%. Only fallow land has increased by 6.49%.

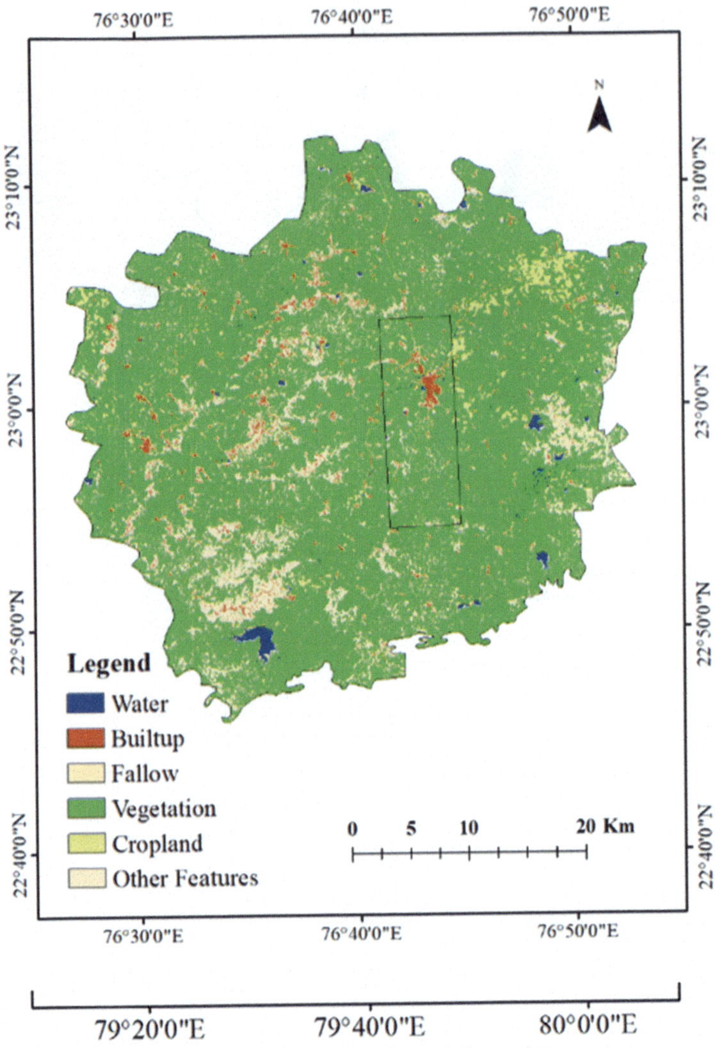

Fig. 7.3 LULC of Bhopal (2000)

In Guwahati, the results show (Table 7.6) that built-up area has increased by 8.1%, whereas the forest cover has decreased by 6.58%, which is due to the impact of urbanization in forest land. The figure clearly shows the replacement of forest due to built-up activities. The reduction in forest land may directly impact carbon sink. Further, fallow land increased by 9.08% and vegetation decreased by 5.97%. The change in vegetation and forest land may directly impact the land use/land cover change as well as the climatic conditions.

The change in land use/land cover from 2000 to 2020 in different regions of India clearly shows the growth of urbanized area. Primarily, the land/use land cover

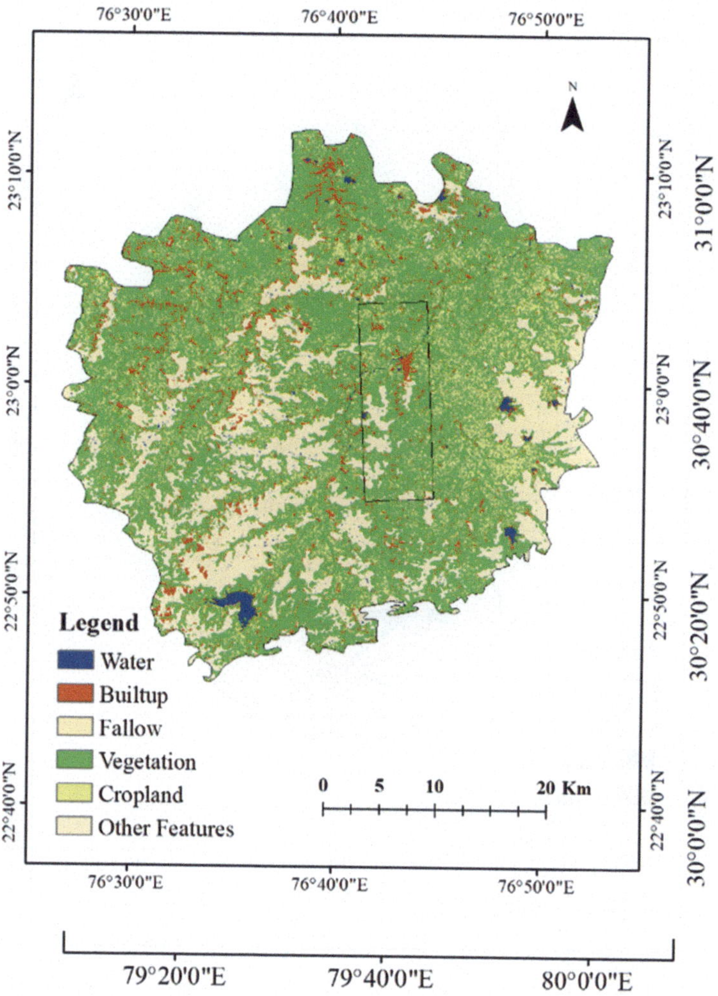

Fig. 7.4 LULC of Bhopal (2020)

Table 7.3 Spatio-temporal change in percentage of area under land use land cover classes (Noida)

Noida	2000		2020	
Features	Urban	Rural	Urban	Rural
Built-up	20.90	9.83	38.29	28.07
Vegetation	34.14	18.26	28.81	25.28
Fallow land	1.74	0.42	8.65	8.22
Cropland	43.22	71.49	24.25	38.43

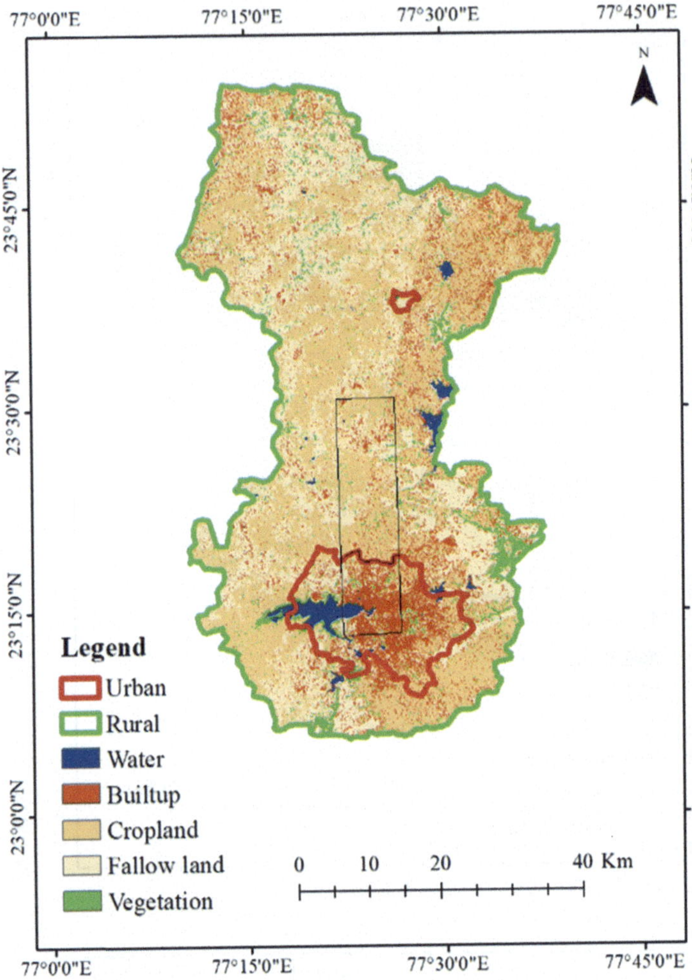

Fig. 7.5 LULC of Noida (2000)

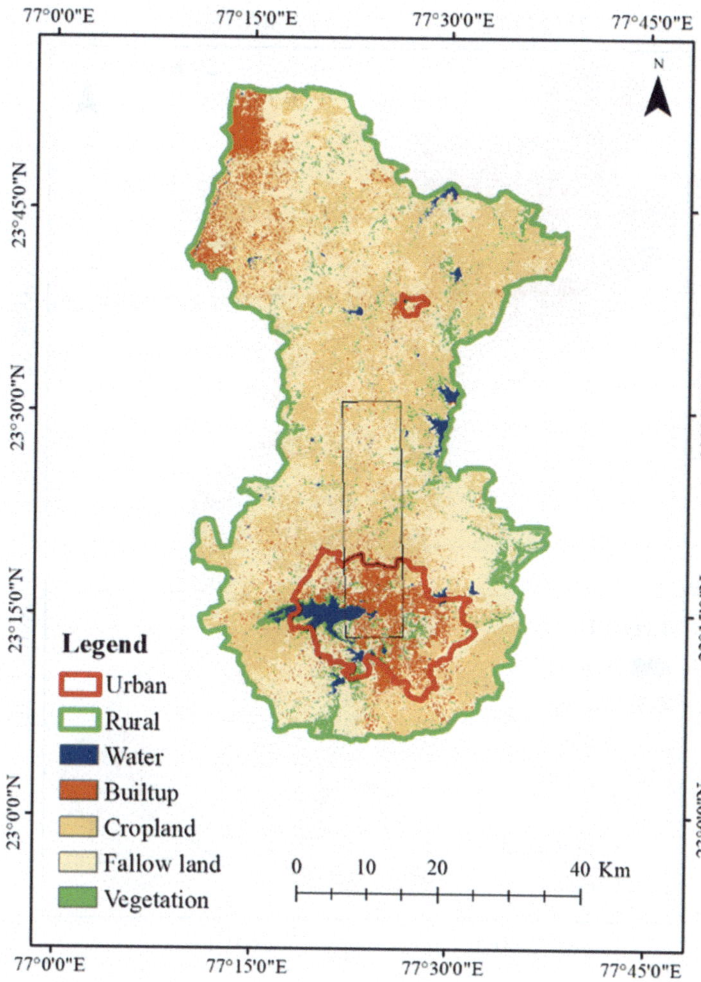

Fig. 7.6 LULC of Noida (2020)

Table 7.4 Spatio-temporal change in percentage of area under land use land cover classes (Ashta)

Features	2000-Rural	2020-Rural
Water bodies	0.63	0.59
Built-up	4.72	8.86
Fallow land	9.08	2.80
Vegetation	78.03	63.39
Cropland	1.63	13.39
Other features	5.91	10.96

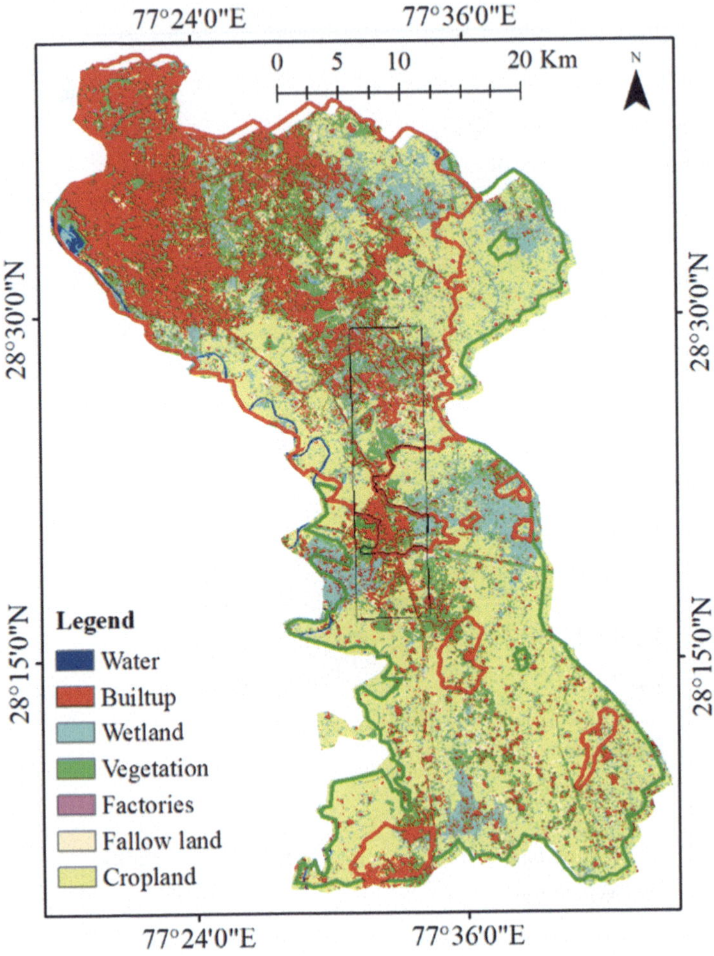

Fig. 7.7 LULC of Ashta (2000)

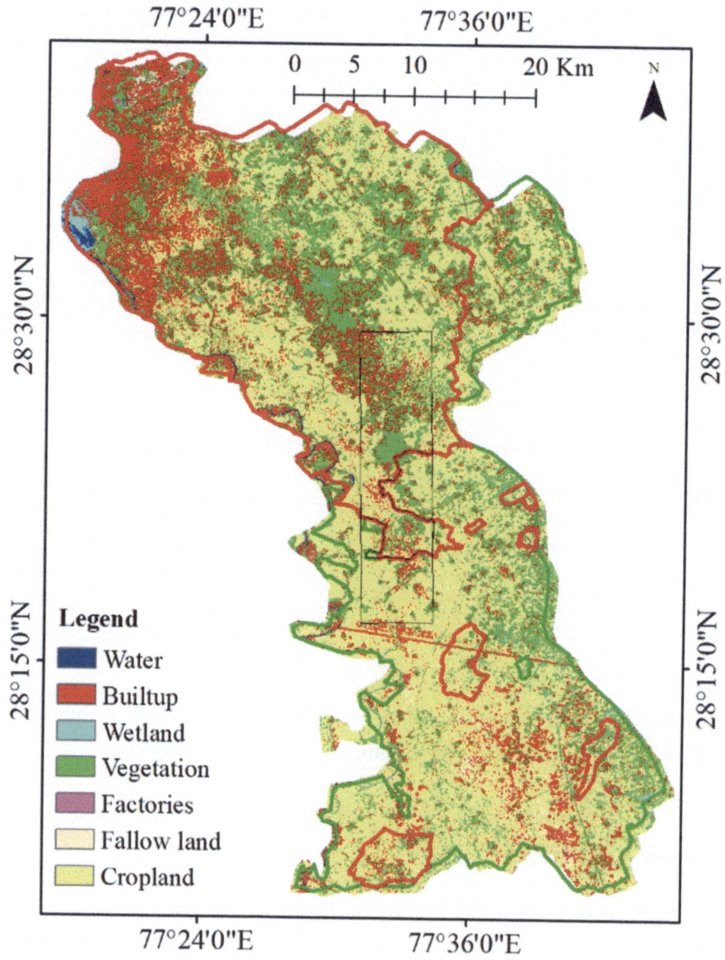

Fig. 7.8 LULU of Ashta (2020)

Table 7.5 Spatio-temporal change in percentage of area under land use land cover classes (Joshimath)

Features	2000-Rural	2020-Rural
Water	2.15	10.00
Dense forest	53.69	33.59
Built-up	2.57	5.89
Ablation area	0.61	0.61
Perforated forest	12.42	11.59
Vegetation	13.07	10.79
Shadow	2.32	7.87
Fallow land	13.17	19.67

Table 7.6 Spatio-temporal change in percentage of area under land use land cover classes (Guwahati)

Feature	2000	2020
Water bodies	0.73	0.75
Built-up	11.27	19.37
Forest	47.19	40.60
Rocky area	13.92	9.28
Fallow	13.12	22.20
Vegetation	13.77	7.80

change is driven by transformation of agricultural land, vegetation land, and fallow land to sprawling urban development setup and industrialization.

7.2 Focus Group Discussion

The report encapsulates the outcomes of the Focus Group Discussion cum Workshop held in Tirupattur (Tamil Nadu) and Bhopal (Madhya Pradesh). The programme was conducted with the participants from different walks of life, occupation, and age group. The programme was organized in four groups. Valuable insights were obtained from the localities and their level of awareness was also documented with the help of questionnaires and extensive sessions. The findings of the discussion depicted that majority of the people are ready to adopt eco-friendly measures to reduce carbon footprint, given that they are provided with assistance and knowledge to do so for sustainable development. *The recommendations from the participants were also sought based on their personal experiences.*

7.2.1 Introduction

Household Carbon Footprint is an important measure to gauge the emission of greenhouse gases that we are accountable for emitting which affects our planet too. Our carbon emission data could help in formulating policies to reduce emissions and provide benefits to the low emitters. Another advantage of knowing our HCF may be to adopt proactive actions for a sustainable and carbon-friendly lifestyle. It can assist to form policies to introduce credit system that will neutralize our carbon usage with the involvement of environmental improvement agencies.

There exists a great rapport between "carbon" and "us." The more unaware we are, the greater the "footprint." As we know, that global warming is potentially a threat to the survival of the living beings; therefore, we need to curb the activities which causes production of greenhouse gases. Our households are also accountable for soaring emissions. According to a rough estimate, households are responsible for adding up to three-quarters to the global carbon emissions. Centre for Ecological Economics and Natural Resources (ISEC, Bengaluru) collaborated with GIZ (New Delhi) to undertake a study on "Household Carbon Footprint in India." It is a part of a bigger project—"Supporting the Institutionalization of Capacities on Climate Change Studies and Actions" implemented by Union Ministry of Environment Forest and Climate Change (MoEF&CC), Government of India and GIZ, New Delhi. This study aims to conduct research on Household Consumption and related environmental impacts in terms of contribution to global climate change. The motive behind the programme was to comprehend people's perception and the motivators that will trigger them to follow eco-friendly path.

To achieve the goal of carbon neutrality, we need to adopt renewable sources of energy. We as a country contribute just over 7% of the global GHG emissions, whereas country like the United States emits more than 14% of the global emissions but China is way ahead of both the counties, contributing approximately 27% alone to the global emissions. Which is more than the USA and India combined. India is working towards the conservation and emission reduction. India has pledged to reduce its carbon sink to 2.5–3bn tons of carbon dioxide equivalent by generating forest covers by 2030 and also to reduce the emission intensity of the GDP by 35% under the Paris Agreement in the year 2016. India is spearheading to tackle the emissions by generating awareness, conserving the ecology and providing amenities to assist citizens to reduce their carbon footprint.

7.2.2 Background

Purpose: The purpose of conducting the "Focus Group Discussion cum Workshop" was to understand people's perception concerning Household Carbon Footprint. To seek suggestions and create a sense of awareness.

Location: The two locations were finalized on the basis of feasibility. The two study areas were also a part of the emissions estimation, they are:

1. Tirupattur (Tamil Nadu)
2. Bhopal (Madhya Pradesh)

Tool used: Perception-selection based questionnaire. The questionnaire was made to capture the thought process of the people, pertaining to their idea of carbon emissions.

Following are the sections of questionnaire:

1. Food and consumables
2. Energy use
3. Transportation
4. Development and environment

Number of Participants: 40 (Tirupattur)

Description: As our study entails assessment of carbon emissions from households, it encouraged us to delve into the mindset of people and their contribution towards the emissions reduction. We have obtained responses considering their behaviour towards carbon emissions and their opinion earlier, but we wanted to get familiar with the nuances of the range of suggestions they had and the lifestyle which they led, which was part of the group discussion, whereas the workshop was to provide them with the knowledge of their daily emissions from various sources which made them to think again and have a rough estimate of how much we add to the global warming. The workshop also helped to spread awareness among people.

The heads of the questionnaire were selected based on the major consumption criteria, i.e., energy use, transportation, food, and environment. The group comprised of participants from different background, social status, education, and age. The first section of the questionnaire that, i.e., "Warm-Up Questions" were the same for all the group where we try to understand the basic idea and information which the participants possess regarding the climate change phenomenon. It incorporated both open-ended and close-ended questions and response was obtained by opinion, standpoint, and show of hands. Also, the "Exit Questions" at the end were also identical for all the three groups. This section encompasses motivation to adopt eco-friendly options, their choice of alternate options, and reward/punishment-based motivators. Placards and bidding cards were used to make the sessions more interactive and elicit response. Four groups were formed under each head and at the end all the groups were merged into one for a common session which is regarded as fourth group.

7.2.3 Group-1: Food and Consumables

In this section, we try to understand and seek information regarding general overview of the participants. To begin with, the respondents were provided with stickers and markers to indicate their views. The next set of questions were "Exploratory" type wherein we sought participants consumption habits of various food types (packed, processed, frozen, imported, organic) and frequency of buying type of food (food delivery or eating outside) and its sources, i.e., local or home grown or from market. We also asked them about their preferences based on availability and price. Questions regarding their willingness to compromise with comfort in order to move towards more eco-friendly lifestyle were put forth. A set of questions under "Exit Questions" were asked at the end.

7.2.4 Group-2: Household Energy (Electricity and Cooking Fuel)

In the Household Energy section, "Exploratory" part included questions based on their idea of type of energy they use and alternate sources which they think can be utilized in their region based on resources. An account of their daily-use appliances was obtained. The interesting part was to ask them about their daily habits which contribute to reduce their footprint, for instance, participants were asked about their habit of turning lights off when not in use, use of CFL, taking short showers, use of low flow taps, etc. Participant's preferences were obtained on their choice of heating water. The waste disposal habits were also the centre of discussion. As a customer, they do check the star ratings (BEE ratings) of their appliances and their willingness to adopt such measure to reduce emissions were also recorded.

7.2.5 Group-3: Travel and Transport

The travel and transport section included questions based on their vehicular possession and their knowledge of the BS ratings of vehicles. Participant's opinion on the policies and regulations based on the eco-friendly transportation was the cornerstone of the discussion. Participants were also asked about fiscal measures, subsidies, and other benefits which they would prefer to be implemented to initiate behavioural changes. Efficient fuel type and preferred mode of transportation were also obtained. We also kept account of the impediments and grievances which held them back from adopting efficient technology. We also provided them with options such as bicycle lane, park, and ride facilities to assess willingness.

7.2.6 Group-4: Development and Environment

All the groups were integrated to discuss potential development areas and how it could benefit environment. The common group session was held at the end to have a holistic overview and opinion of the people regarding India's means and ways to become carbon neutral. The area in which the participants reside was taken as base for ensuing questions. A discussion on infrastructure which is imperative for the local people was taken into consideration too. Opinion on measure by the government which could boost standard of living was noted. A series of cost and benefits of different scenario was also discussed with the participants.

7.2.7 Conclusion

Participants were fairly aware of the climate change phenomenon regardless of their distinctive backgrounds. They were proactively responding the environmental-friendly measure which was discussed and they have adopted in their lifestyle.

On the basis of the discussion for all the groups, it can be summed up in the following heads:

Food and Consumables—Majority of the participants of this particular group said the climate change phenomenon is a hoax. Avoiding plastic and planting trees constitutes a part of their daily eco-friendly practices, and they believe that little changes in habit could avert the current climatic conditions.

Consumption of meat is particularly high for this particular region and majority of the food grown are local, but their food habits having been changing with the time, the taste of the food has been changed in the past 30 years, asserted the senior citizens. The reason behind changing food options is majorly attributed to health concerns and migration for work, such conditions have changed food habit drastically and consumption of fast food is on the rise. According to the local people, advertisements and the taste of variety of food have made them to turn away from locally grown food and order or eat outside. It was alarming to learn that people were not ready to get out of their comfort zone and make changes in their choice of cleansing agents and use of paper to become eco-friendly. In the list of type food preferences, regardless of their price, organic food stood at the top followed by home grown and then processed food. All participants asserted that the dearth of information on the eco-friendly alternatives is an impediment to adoption of positive measures. Majority was in favour of spending not more than 10% of their income for adopting such measures.

Household Energy—Participants were well versed with the naturogenic and anthropogenic activities that majorly contributors to climate change, for instance, deforestation, radiation, urbanization, effluent, e-waste, etc. They believe that the effect of climate change is immense on the human health, but the effects are sometimes overemphasized. It was surprising to know that the people do have an idea of the CFCs used in their refrigerators and air conditioners. Some of the participants used cow dung as manures and "Gober Gas" for cooking purposes. They also had small garden in their home to grow their own food. The visible impact of climate change in their area included agricultural floods during harvest period, pest attack, and soil erosion.

Major sources of energy, which are used frequently by them are firewood and bio-energy. The fridge, electric heater, water pump, and washing machine were frequently used appliances in the households. Majority of the participants are inclined towards behavioural change instead of buying energy-efficient equipment. They do maintain high willingness to adopt affordable alternatives to reduce emissions from energy use. When it comes to buying energy-efficient appliances, participants would be happy to buy efficient appliance provided that they are informed about the product by the supplier or through eco-labelling. For this particular group, subsidies on the appliance could help them become more energy efficient.

Travel and Transport—In this particular group, people generally equated climate change with natural calamities, e.g. flood and excess heat. Use of public transport and saving water was the two major options that they practiced in their day-to-day life to reduce emissions. The participants were aware that the consequences of climate change will be dealt by the future generation, but they believed that the environmental impacts are frequently overstated. Participants are aware of the emissions from the factories and impact of methane on the environment. For this group, the effects of climate change experienced by them include change in water table, crop damage, rise in temperature, and salinization of soil.

The most possessed vehicle in this particular area is bike, i.e., two per household, but they are not much aware of the BS ratings, hence they do not ask about it while purchasing vehicle. The majority of the participants are in favour of availing public transport, and hence they demand an overhaul of the current transport system wherein better connectivity of the transport system could play a major role also, they will not mind if the price of the petroleum is increased. According to them, behavioural change, awareness, and positive tax benefits would act as an impetus for the people to avail more of public transport. Participants voted electric vehicles as most environment-friendly fuel-run vehicles and diesel vehicles as the least environment friendly. According to them, political influence is the most influential factor which would affect introduction of new fuel vehicles type. The motivating factor for them is encouragement from friends, society, and community, and they are ready to compromise with the quality if subsidies are provided to them to adopt alternatives for emission reduction.

Development and Environment—The common group had the maximum number of participants. Here participants were asked about the infrastructural developments for the people which is crucial under different heads. The rankings are as follows (most important to least important):

1. Physical infrastructure (Roads, water supply, wastewater treatment, waste management, bridges)
2. Conservation and restoration of land and water resources
3. Health infrastructure
4. Livelihood generation
5. Educational infrastructure (govt. schools)
6. Conservation of biodiversity

A discussion regarding the factors which provokes people to pollute included lack of awareness as the most selected option followed by minimum government support, societal construct, behavioural change, and lack of facilities. Few of the participants expressed concerns about the development having effect on the environment but were countered by the fact that if the development is administered diligently, keeping in mind, the adverse effects on environment is kept minimal and are compensated, it could prove to be effective.

7.2.8 Recommendations

The recommendation mentioned are purely based on the response of the participants, they are as follows:

- To initiate awareness through creative means is necessary to inform people about the climate changes which are imminent.
- Enforcement of prohibitory measure regarding usage of age-old appliances and vehicles to reduce emissions.
- Reduction in price of good quality of pesticides and promotion of nature-based traditional methods of removing pests.
- Introducing models at panchayat and village level to demonstrate usage and benefits of low-emission equipment.
- Involving Non-Governmental Organizations (NGO) to demonstrate use and effect of various technological advancements for emissions reduction.
- Targeting special groups who are unaware and have high emissions percentage.
- Organization of community level meetings.
- Provide incentive on the equipment which has low impact on environment as people are ready to compensate a bit with the quality.
- Renewable energy sources should be supplied in efficient manner and on a large scale.

Case Studies

8

8.1 Case Study-1: Western Himalayas—A Spate of Emissions

8.1.1 Highlights

- Series of tragic incidents in Western Himalayas attributed to warming of the atmosphere.
- Use of dual sources of household energy by the residents of Chamoli district, Uttarakhand.
- Highest emission figures in household energy sector from Joshimath in comparison to all the selected regions.
- Proportional relationship between the living area and the HCF.

8.1.2 Introduction

It is now legible; the human habitat and its environs are feeling the climatic thump in Himalayan region. The winters are turning out to be drier and the summers wetter though not exceptionally. Factoring out the Polar Regions, the Himalayas embraces the largest cryosphere and bestow millions with its resources. One of the consumers of the resources are the residents of Joshimath also known as Jyotirmath. A scared place located at the height of 6150 feet within Chamoli district in Uttarakhand. The gateway to the scintillating mountain range holds 3898 households with the population of 16,709 according to the 2011 census, which continues to burgeon undeniably. Uttarakhand is situated in the Western Himalayas which holds 15% of total ice mass of the entire Himalayas; it has been taking up the heat, literally. Glacial Lake Outburst Flood (GLOF), cloud burst, flash flood, and the concomitant delude has been tearing down the place lately. We can put onus of the catastrophe on the climate change, but honestly, we know who is guilty of it. To that effect, we tried to

S. Nautiyal et al., *Determinants and Stratification of Household Carbon Footprint*,
https://doi.org/10.1007/978-981-96-2739-4_8

limit our focus to carbon emissions which is the mainstay of the research. The emissions are the prime reason for receding glaciers in the Himalayan region. While the authorities have pledged to stick to accords and decadal targets to increase carbon sink and cut down emissions, it can be effectively thwarted at personal and household level as well. The key is change in lifestyle, knowledge, and technical know-how.

8.1.3 The Incident

Turning a blind eye to the rising climatic concerns will surely render the situation irremediable. The recent glacial disaster in Chamoli district is one such harbinger. On February 7, Joshimath encounter a tragic flood triggered by glacial outburst which led to loss of lives, properties and also destroyed two hydro-power projects. More than 80 people lost their lives and many went missing. Another glacial outburst on 24th April startled the habitants of the Niti Valley again, which is near Joshimath, fortunately, no loss of life was reported. Sporadic cloudbursts have a history with Uttarakhand, the latest incident occurring on 11th May. Rudraprayag, Uttarkashi, Tehri Garhwal, and Chamoli have been hampered recently, by the phenomenon. The reason is being attributed to one of the conjectures which is the steady warmer months in Uttarakhand. Currently, though the availability of data is scarce to validate the causes of the aforementioned incidents but what we can comprehend, prima facie, is that the major benefactor to these climatic disasters is the warming up of our earth.

8.1.4 The Cause

The motive behind the instantiation of the recent calamities is to inform about the severity of the climate change and to direct your attention to the source of the corollary. The data extracted for Joshimath (Uttarakhand), which constitutes a portion of the study area in the Western Himalayas have provided us with a deviant behavioural aspect concerning household carbon footprint. The energy demand of the population is met by the Hydro-Power Plant situated in the Chamoli district. Rural areas were taken into account to calibrate the quantity of emissions as they have multiple alternatives when it comes to energy usage. Sundry schemes have been rolled out for the electricity access to households such as Deen Dayal Upadhyaya Gram Jyoti Yojana (DDUGJY) and Integrated Power Development Scheme (IPDS). Bearing in mind the rate of energy use, Joshimath has shown the highest carbon emissions in the energy category (55%), overall.

The average carbon emissions from electricity consumption of households in the rural areas of Joshimath stood at 840.64 kgCO_2e/year, whereas the emissions from the biomass usage stood at 1790.41 kgCO_2e/year which, by contrast, is the highest among Eastern Himalayas, Middle Gangetic Plains, Eastern Plateau and Hills, Southern Plateau and Hills, and the Central Plateau and Hills (Fig. 8.1). The

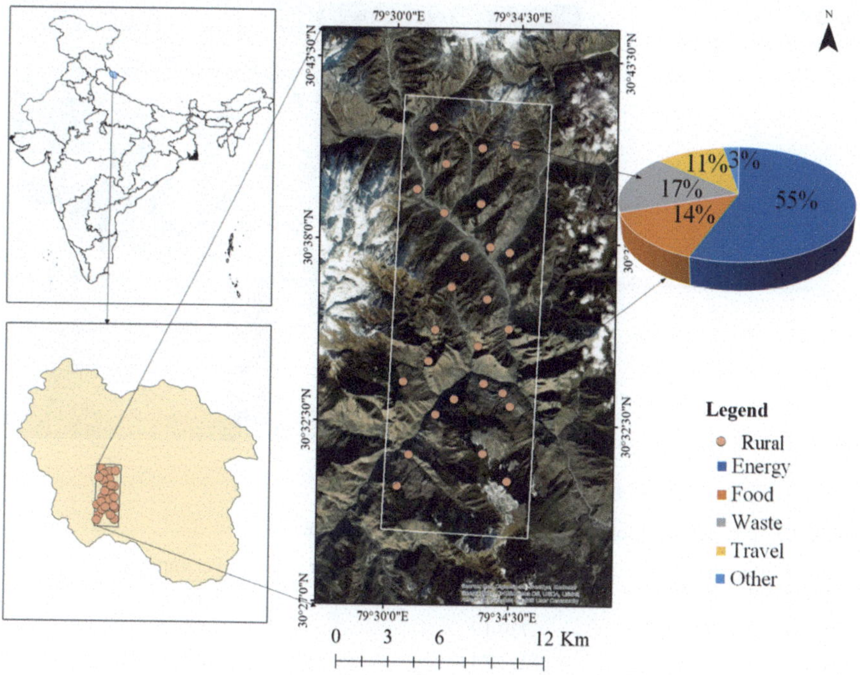

Fig. 8.1 The map represents the rural settlement in the Joshimath region

organic carbon exudes from the high usage of biomass which is sufficiently available.

Petrogenic sources too play role in the emission and partial combustions of non-renewable fuels have been greatly held accountable by some researchers. According to NITI Aayog report on Sustainable Tourism in Indian Himalayan Region, the Western Himalayas have elevated tourist influx than entire North Eastern region combined. It has opened a source of income for the rural residents and as reported by the Uttarakhand's Tourism Department, the number of rooms with facilities is being constructed as the inhabitants are keen on entertaining hosts as it provides comfortable income to them. The analyses suggest that the built-up area in the rural areas commensurate with the total carbon emissions per year (Fig. 8.2).

A correlation has also been observed between high emissions and people holding land up to 1 hectare (Table 8.1). This trend has been observed in the rural areas only and not in the urban zone within Chamoli district. As a matter of fact, the 2017 data suggest that the accommodation for the tourist is exclusively built in the rural settings only. It not only catapults the carbon emissions to a higher degree from households for energy usage, increased appliances, traditional methods of energy consumption in the absence of the factitious resources, but also exerts pressure on the natural resources.

Fig. 8.2 Residents carrying fuelwood

Table 8.1 Estimation of total HCF with respect to built-up area

Built-up area (in sq. ft)	Total emissions kg/HH/year	Per capita CF
<500	8709.83	2.17
500–800	6739.12	1.68
>800	25472.84	6.36

8.1.5 Resolution

On the whole, potential implications of the climate change on the Western Himalayas require more deliberation both on the research and policy front. In the case of climatic phenomenon, meteorological data and instrumentation to detect climatic variability and extreme climatic events is necessary. Research on the proportion of air pollutants (GHGs) considering both organic and inorganic carbon emissions in the Himalayan atmosphere needs to be administered. The periodic forest fires in Uttarakhand have been affecting multi-hectares of forest land both in the reserved and civil areas which is again ascribed to dry weather and high fuel load. The forest fires now are being observed every year hence the authorities have to be prompt in availing the requisite equipment and respond rapidly to stop such incidents. Responsible tourists should only be encouraged and the residents should make sure that the element of green infrastructure and judicious use of resources becomes a part of their lifestyle. Best practices among the local bodies needs to be promoted with befitting aid. Corroborations regarding the utility of traditional knowledge of variations in the climate and mitigation strategies should be tested to turn down the climate change impact.

8.2 Case Study-2: Tirupattur—Thinking Outside the Scrap

8.2.1 Highlights

- Lowest carbon footprint from household waste.
- Distinctive waste management practices and reward system by the authority.
- Circular flow practices by the self-help groups in Tirupattur.

8.2.2 Introduction

In the Southern Plateau and Hills lies a city of sandalwood known as Tirupattur in Tamil Nadu. The expanse of sandalwood trees in the Yelagiri and Jawadhu Hills is the cause of the bestowed sobriquet. The total population of the place is 83,612 as per the previous census. Our study was carried out to comprehend the emissions statistics of the households, individual perceptions, and most importantly their actions towards carbon emissions. It was fascinating to observe that a minor population do have an idea about the climate change phenomena and are working towards it, whereas some dismiss the looming threat as a hoax, though the people of Tamil Nadu are not unmarred by the climatic fury. They have been overwhelmed by the loss of life and property owing to incessant rainfall within the urban areas and inundation in the outskirts. Cyclone Nivar being the recent addition to the list of catastrophes, the natural disasters have made it difficult for the people to rebuild their lives time and again. The yearly flooding in several parts of the state prompt other cascading effects which lead to health epidemic congested waste deposits and so on. The sufferance is up on us because what goes around comes around, not to mention the role of the residents, political authorities, public and private entities.

8.2.3 The Human Hand

The very basics of ecology has been disobeyed. Within the capital, the airports have been built on the floodplains of the river Adyar, and rampant construction on acres of land and parts of marshlands, which acts as a sink for the flood water, should be avoided at all costs. Rapid growth of infrastructure and urbanization has stupefied the municipal authorities. According the 2018 figures provided by Tamil Nadu government, they are far ahead on several states in the waste management sector and has proved to be 95% efficient in waste accumulation, with 67% efficiency in door-to-door collection. But there's a snag too, about 90% of the total waste is left untreated and 98% of it is deposited in the open space which poses a serious threat to environment and public health. About 33% of the residents feel free to dump their waste erratically on the roads, streets, neighbouring plots and drains devaluing the environmental well-being. Now, the issues are not limited to behavioural change, it is more of a managerial one. The change in lifestyles demands adequate planning and budgeting of the urban, peri-urban, and rural zones. The waste management services need to be ramped up. The unwavering deficit of manpower, in-house capabilities, and public awareness are perpetual impediments. There is a silver lining in the entire setup too, the participation of private entities and their efforts to gather recyclable material from the scrap and sell it to the industries which constitutes a part of the circular economy. But some of these materials are dumped into drain which clogs the system and some wind-up at the dump sites which again is dealt by ragpickers.

8.2.4 The Emissions

The major benefactors of GHG emissions in the given areas are excessive use of private vehicles as the public transport is frail, reliance on grid energy, and waste mismanagement. The developments are on the rise in the wind and solar sectors and introduction of provisions to ensure more public transport such as the recent exemption of women for fare-free bus transit by the Government of Tamil Nadu, with introduction of zero emissions bike may help in emissions reduction (Fig. 8.3).

To our surprise, we found out that residents of Tirupattur (Fig. 8.4) who were the participants of the study are very much fascinated with the waste management practices and cleanliness.

Waste has been considered for the case study which is contributing very less quantity of emissions among other sectors within households of Tirupattur (Fig. 8.5). The analysis suggests that the emissions from the waste from Tirupattur, in contrast to other selected areas, are mere 3% of the total emissions.

The significant low waste to emissions ratio is a direct result of favourable practices. The segregation of the waste is sometimes done at source and also by the municipal waste collectors. A sum of money as tax is also paid by the residents in panchayat office for better waste-related services. Furthermore, the residents are willing to pay 10% of their total income towards emissions reduction. In the rural zone, varying self-help groups have taken up the initiative to make the environment conducive to life. The SHGs have set up a mechanism to make use of the bamboo and carve articles out of it. They are producing jewellery, ornaments, embellishments, utensils, and other decorative materials that have aided them financially and

Fig. 8.3 Tirupattur bus stand

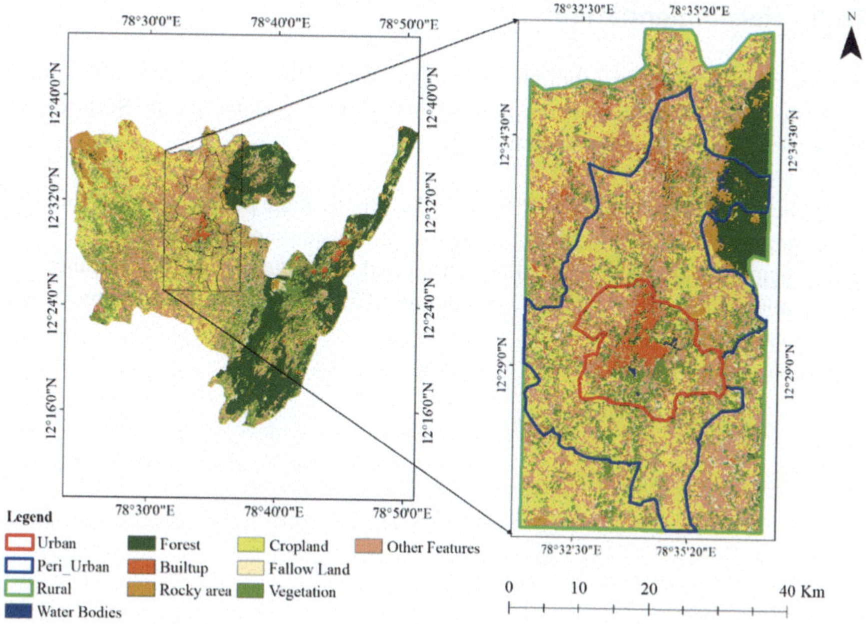

Fig. 8.4 Location map of Tirupattur

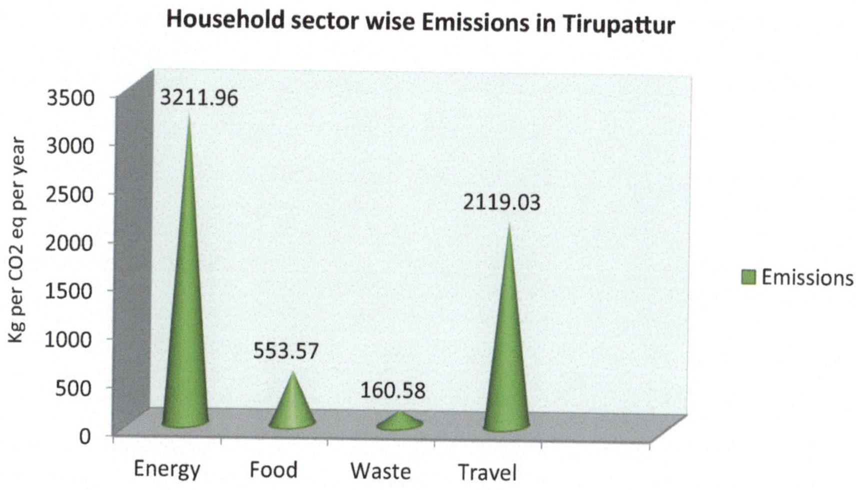

Fig. 8.5 Household sector-wise emissions in Tirupattur

contributed to lessen the emissions to some extent. After making the use of such material, they are disposed off by composting. The efforts by the municipality need to be addressed too as they explicitly reward those residents who follow 4Rs and adhere to good waste management practices.

8.2.5 Recommendations

The drawbacks still persist though; in some pockets of the area, people still incinerate waste material and illegal dumping of waste still exists in Tirupattur. Some of the recommendations from the case study are:

1. Public-Private Partnership (PPP) can be helpful to make people aware of impacts of improper disposal, recycling of waste, and waste management.
2. Municipality should periodically monitor and eradicate the illegal dumping.
3. Collaboration of Government, NGOs, and Community groups to achieve the concept of zero waste.
4. Polluter pays principle should also come into effect.

9 Impact of COVID-19 on Consumption Pattern of Indian Households

The outbreak of Covid-19 has caused substantial economic crisis and recession across the globe due to social distancing measures (Martin 2021). These social distancing measures (self-isolation and travel restrictions) interrupted all the economic activities and resulted in reduced workforce in all the sectors (Nicola et al. 2020). Social distancing measures not only resulted in decreased economic activities, it has also affected mental, physical, social, and spiritual health of the people across the globe (Poudel and Subedi 2020). The degree of impact (economic, social, spiritual, etc.) due to restrictions is different in various income groups among urban, peri-urban, and rural areas.

9.1 Impact on Economy

People's perception of economic loss among all income groups shows that maximum impact of reduced economic activity was found to be higher in rural areas followed by peri-urban, and urban areas. In rural areas, total of 50.47% of respondents said that Covid-19 pandemic had negative impact on economy of their households. Approximately 25.97% people from rural areas lost their jobs which is the major reason for their economic deprivation. Similarly, 42.46% of respondents from peri-urban areas and 31% of respondents from urban areas witnessed economic loss due to the Covid-19 restrictions.

Among the six income groups, negative implications of the economy were minimal upon the higher income classes, when compared to other income categories in the three zones (Table 9.1). However, in peri-urban zones, high-income category (>10 lakh) respondents (50%) agreed on the losses they had to encounter due to economic downturn. However, there is no impact on the economy of high-income group of rural zones. Among all income groups in rural zones, high economic impact was observed upon the medium income groups whereas large income group had to bear no economic impact at all. Similarly, in peri-urban areas all income

S. Nautiyal et al., *Determinants and Stratification of Household Carbon Footprint*,
https://doi.org/10.1007/978-981-96-2739-4_9

Table 9.1 Impact on economy on rural, peri-urban, and urban areas among different income groups

	Rural (%)		Peri-urban (%)		Urban (%)	
Income groups	Yes	No	Yes	No	Yes	No
<50 thousand	45.7516	54.2484	47.3684	47.3684	52.9412	47.0588
50 thousand to 1 lakh	47.8161	52.1839	47.619	52.381	48.7179	51.2821
1 to 3 lakhs	58.5185	41.1111	45.8599	54.1401	33.3333	66.6667
3 lakhs to 5 lakhs	45.5882	54.4118	33.3333	66.6667	21.9697	78.0303
5 lakhs to 10 lakhs	55	45	37.2549	62.7451	21.0526	78.9474
>10 lakhs	0	100	50	50	5.88235	94.1176
Total	**50.4752**	**49.4192**	**42.4691**	**57.284**	**31.0112**	**68.9888**

Table 9.2 Impact on jobs in rural, peri-urban, and urban areas among different income groups

	Rural (%)		Peri-urban (%)		Urban (%)	
Income groups	Yes	No	Yes	No	Yes	No
<50 thousand	18.3007	81.6993	0	100	5.88235	94.1176
50 thousand to 1 lakh	27.3563	72.6437	20.2381	79.7619	16.6667	83.3333
1 to 3 lakhs	30	69.6296	18.4713	80.8917	15.7407	84.2593
3 lakhs to 5 lakhs	20.5882	79.4118	13.3333	86.6667	9.84848	90.1515
5 lakhs to 10 lakhs	20	80	11.7647	88.2353	6.57895	93.4211
>10 lakhs	0	100	0	100	0	100
Total	**25.9768**	**73.9176**	**15.8025**	**83.9506**	**11.236**	**88.764**

classes had to face the economic impact due to Covid-19. However, in urban areas lower income groups observed high economic impact, high-income groups observed low economic impact. The following table exhibits impact of Covid-19 on the economy in rural, peri-urban, and urban zones.

Many people across the country lost their jobs due to the shutting down of all the financial activities, both in organized and unorganized sectors in the view of Covid-19 pandemic situation. Job loss scenario was observed in rural, peri-urban, and urban areas among all the economic groups. The analysis results concerning jobs (Table 9.2) revealed that 25.97% of people from rural areas admitted that they have lost their jobs during Covid pandemic situations, similarly 15.8% of people in peri-urban zones and 11.23% of in urban zones lost their jobs. Among different economic groups, the rural, peri-urban and urban zones, high-income classes (>10 lakh) did not face any job loss, as most of them are working in organized sectors. However, people from middle-income groups (50 thousand to 5 lakh) lost their jobs. These middle-income group people are mostly employed in informal sectors. According to the ILO report, informal sectors are highly vulnerable during Covid pandemic situations. This could be the reason for the loss of jobs of the middle-income groups. However, low-income classes (<50 thousand) from the three regions were less vulnerable to losing their jobs. The detailed analysis on impact on jobs in rural, peri-urban, and urban areas are depicted in the following table.

At present, there is no appropriate mechanism to predict the worldwide impact of Covid-19 on the world economy, but economist have come to a consensus that the major economies will lose, at least, about 2.9% of their GDP when compared to the previous year 2020. The losses are largely driven by demand for goods and services which has fallen drastically. From tourism to airline industries and from oil prices to automobile industries have been experiencing steep downfall and need immediately governmental interventions to stabilize losses to some extent. Lay-offs by several organizations have created a sense of despair among the people. Both the organized and unorganized sectors have been hit hard due to loss of revenue. The distress among the upcoming workforce is apparent. The economists have begun to anticipate possible recession.

India is amidst second wave of Covid-19. It was on the path of recovery which was visible from the third quarter of the financial year 2020–21. Though the GST collection has been fairly above the one trillion mark and new addition to the list of start-ups have been observed, the employment sectors remain ominous. In May 2020, the unemployment rate was 24% then it slumped to 7% in the beginning of the year and now when the second wave has hit the figure stands at 9% according to the Centre for Monitoring Indian Economic (CMIE) statistics (Table 9.3). Unemployment rate as of April 2021 for the selected study area is given below:

Table 9.3 Unemployment rate of the states

S.no	States	Unemployment rate (%)
1.	Assam	0.2
2.	Jharkhand	16.5
3.	Karnataka	2.0
4.	Madhya Pradesh	1.4
5.	Tamil Nadu	2.3
6.	Uttar Pradesh	6.3
7.	Uttarakhand	6.0

Source: http://unemploymentinindia.cmie.com/

9.2 Impact on Employment Among Households

The current analysis is concerned with the first wave of the Covid-19, as the data collection took place in the year 2020. The following chart represents the economic impact of Covid-19 on the households of the selected region. The CMIE data suggest that the current unemployment rate of the urban population is 10.9%, whereas it is 9.4% for the rural population. The inputs obtained from the households shows that the economically poor households in the urban setting are highly affected by the pandemic in terms of income, facilities, and nutrition. Communities and the Public Distribution System (PDS) have provided some relief to the habitants. The rural poor have also been provided with direct cash transfer and exemption for buying ration among other benefits.

The data provides (Fig. 9.1) an insight into the variation of impact on the households in different settings. About 53.9% of the respondents in the rural part of the selected study area have admitted that their financial condition has been hampered

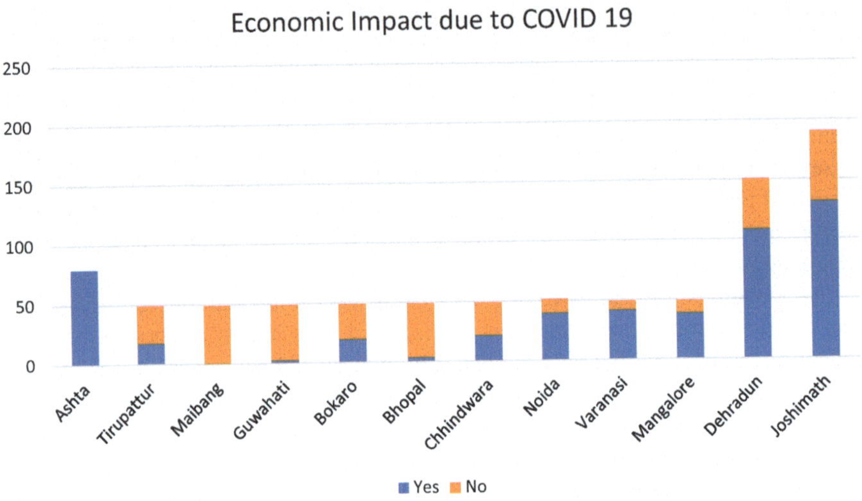

Fig. 9.1 Economic impact of Covid-19 on the rural households

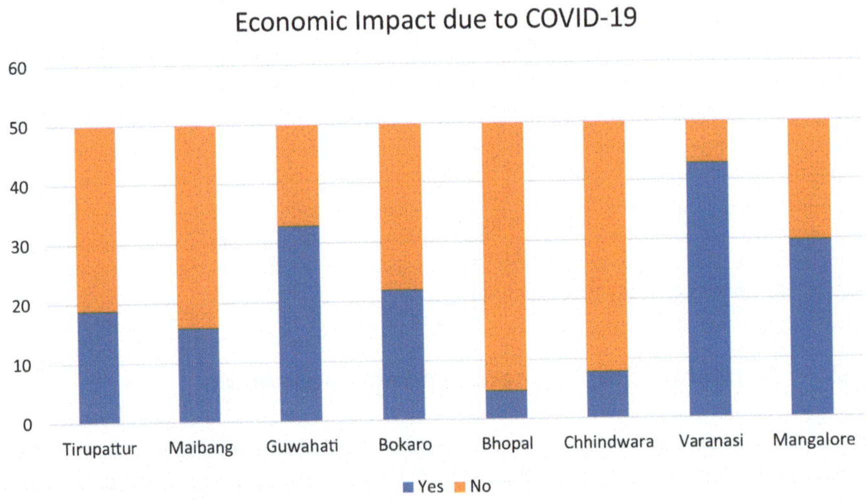

Fig. 9.2 Economic impact of Covid-19 on the peri-urban households

by the first wave and lockdowns. About 30% of the respondents have experienced job loss.

The above Fig. 9.2 suggests that about 44% of the peri-urban households have been affected economically which can be attributed to their better savings and sound knowledge about financial management. About 19.75% of the households have been experiencing job loss.

About 54.18% of urban households have mentioned that they have been affected by the Covid-19, which is higher than the rural households. But the extent of impact

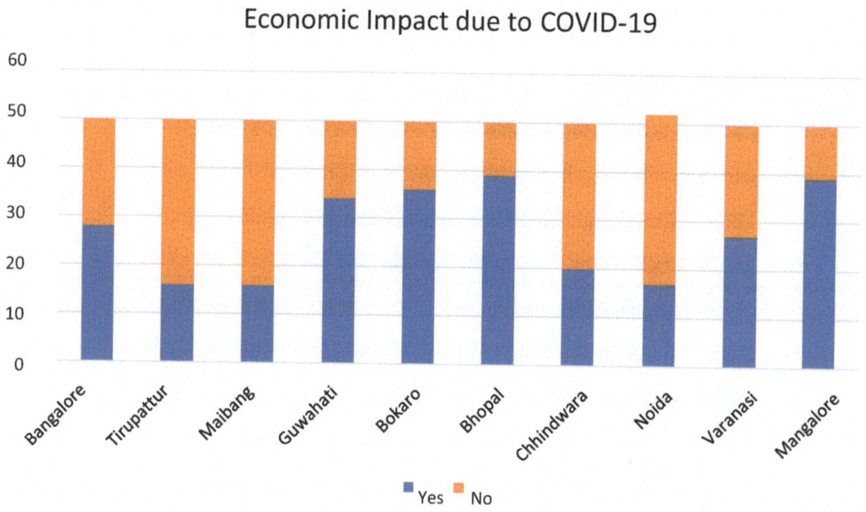

Fig. 9.3 Economic impact of Covid-19 on urban households

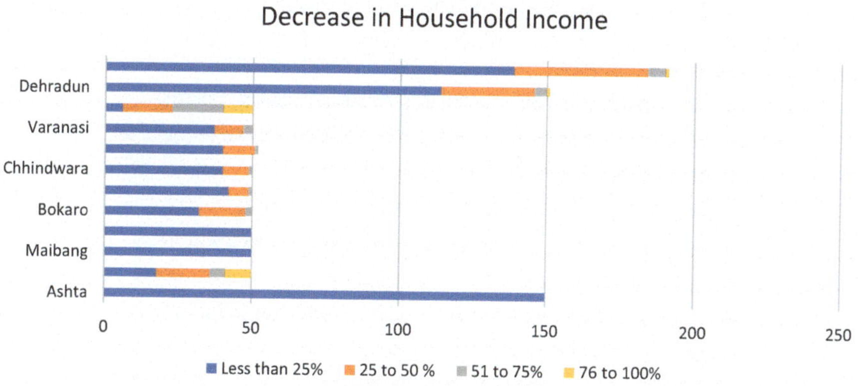

Fig. 9.4 Percentage decrease in the income of rural households

is not as severe as compared to the rural and peri-urban areas. About 19.75% of the households have experienced job losses (Fig. 9.3).

9.3 Impact on Income

From the rural household data (Fig. 9.4), it was concluded that about 76% of the rural households have experienced 25% or less reduction in their total income of the households followed by 17.4% has experienced reduction of 25–50%, whereas 4.23% has experienced 51–75% in the income and 2.22% has experienced 76–100% loss of income.

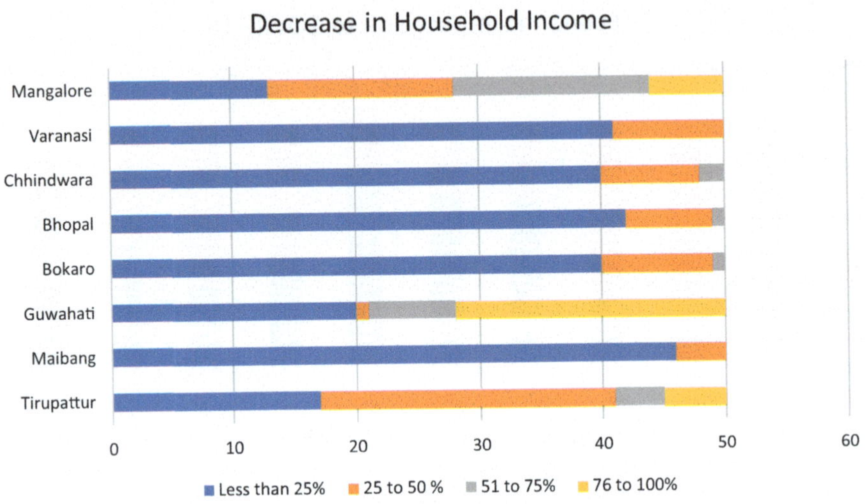

Fig. 9.5 Percentage decrease in income of the peri-urban households

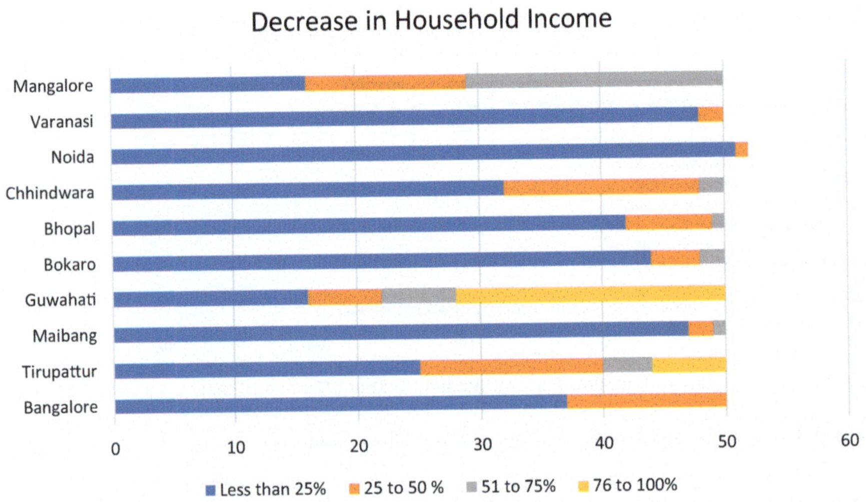

Fig. 9.6 Percentage decrease in income of the urban households

In the peri-urban households (Fig. 9.5) around 64.75% of the households have experienced downfall of 25% or less in their total income, 19.25% of the households have experienced 25–50% decrease, 7.75% of households have experienced 51–75% decrease, and 8.25% households has experienced 8.25% of decrease in total income.

The urban households (Fig. 9.6) have experienced about 71.31% of decrease in their total household income followed by 15.73% households experienced 25–50% reduction, 7.37% households experience 51–75% reduction, and 5.57% of households have experienced a reduction of 76–100% in their total household income.

9.4 Usage of Public Transport

Usage of public transport was reduced all across the country during and after Covid-19 pandemic. Maximum number of respondents from the three zones (rural, peri-urban, and urban) asserted to (more than 70%) discontinuation of public transport usage. People using public transport during Covid-19 were exiguous. The detailed analysis on usage of transport in rural, peri-urban, and urban area are depicted in the following Table 9.4.

The study entails that the energy usage rose during the series of lockdowns. 4% of rural households belonging to low-income groups only reported decrease in electricity consumption; whereas, 44% of the households reported an increase in electricity consumption across all the zones and income groups (Table 9.5). As people had to stay in the house and work from home elevated energy consumption was inevitable in urban and peri-urban areas. In rural areas, people resorted to going back home and search for income alternatives. Household cooking added to extensive use of LPG, induction, and other forms of cooking fuel in urban and peri-urban areas whereas biomass energy was utilized more in rural areas for cooking; sporadic rise in LPG prices also affected the rural households. An increase in entertainment, online interaction, videos, and other leisure activities are also responsible for increased energy use. Daily household needs in changing season has to be met hence, use of AC, coolers, heaters, laptops, phones, and other electrical appliances have added to the excess and longer period of electricity consumption by households.

Emissions by transportation are largely attributed to cessation of public transport and frequent use of private travel. When the lockdown was imposed, transportation of any sorts were minimum but with easing of the lockdown people started to travel to places in their personal cars and bikes, either for commuting or leisure purposes, though emission level remained low when compared to pre-Covid times.

The analysis entails a probe into the food consumption pattern of the households before and during the lockdown induced by Covid-19 pandemic. Overall purchase of groceries is still on the rise. The current study has revealed the consumption of various food items with respect to income and occupation. High difference in the variation of consumption can be observed in the higher income classes in contrast to lower income classes.

The following analysis entails food consumption based on the type of primary occupation of the households:

The primary occupation of the head of the households has been taken into consideration. Table 9.6 represents occupation-wise food consumption which suggests that the private sector employees have the highest consumption of cereals and is at par with the agriculture and allied sector workers, when it comes to consumption of pulses. The private sector employees have higher consumption of milk and milk products, vegetables, fruits, chicken, fish, and meat whereas the workers in the NTFP sector have lowest consumption of all the food materials. The results suggest that the private sector employees, who are largely comprised of high-income group in our study, have higher increased consumption in all categories of food except chicken; the personnel employed in government sector have high increased consumption of non-vegetarian food.

Table 9.4 Usage of public transport in rural, peri-urban, and urban areas among different income groups

Income groups	Rural (%)			Peri-urban (%)			Urban (%)		
	Not using	Reduced	Same as before	Not using	Reduced	Same as before	Not using	Reduced	Same as before
<50 thousand	57.52	38.56	3.92157	73.6842	21.0526	0	55.8824	38.2353	5.88235
50 thousand to 1 lakh	77.01	21.84	1.14943	86.9048	13.0952	0	78.2051	17.9487	3.84615
1 to 3 lakhs	70.74	26.3	2.96296	73.8854	22.9299	3.18471	69.4444	26.8519	3.7037
3 lakhs to 5 lakhs	76.47	23.53	0	78.8889	21.1111	0	74.2424	19.697	6.06061
5 lakhs to 10 lakhs	100	0	0	76.4706	23.5294	0	61.8421	30.2632	7.89474
>10 lakhs	100	0	0	100	0	0	88.2353	11.7647	0
Total	**72.54**	**25.45**	**2.01**	**78.27**	**20.25**	**1.23**	**77.78**	**26.42**	**5.68**

Table 9.5 Distribution of households with increased electricity consumption during Covid-19 and corresponding average increase in emission per household

	% of HH reporting increase in electricity consumption			Increase in CF from electricity $KgCO_2e$ per HH per year		
Income groups	Urban	Peri-urban	Rural	Urban	Peri-urban	Rural
Very low income	33.94	23.45	57.05	336.60	265.20	149.02
Low income	22.42	39.31	32.73	309.98	274.38	191.45
Middle income	19.39	24.83	7.05	705.64	681.56	180.44
High income	24.24	12.41	3.18	759.29	833.65	494.39

9.5 Discussion

Since the pronouncement of lockdown on the eve of 24th March 2020 due to the Coronavirus scare, the adverse effects on economy, food, and agricultural system were conspicuous. As the essentials were only allowed to be sold, the increased procurement of packaged food, rice, flour, oil, eggs, milk products, and other eatables was observed. The focus shifted towards home-cooked food. The anxiety related to food was not only observed in India but other countries too, such as Denmark, Germany, and Slovenia; these countries too faced decline in income due to Covid-19 (Janssen et al. 2021). The consumption frequency has been higher all over the world at differentiated quantities. In some countries, such as Qatar, healthy consumption was observed, where people started drinking more water and consume more fruits (Ben Hassan et al. 2020). Majority of population started to differentiate between what to consume and what not to consume.

The analysis is largely based on the socio-demographic sample composition and the impact of Covid-19 in the behaviour of the consumption frequency of different households. A decline in the consumption of non-vegetarian food was observed; initially, a link was being established between meat consumption and the virus, without substantiation. Due to the migration of workers from the agricultural fields, in the later period, affect harvesting of the agricultural produce. As we know that dinning services were entirely shut across the nation, the delivery services were available sparingly and mostly limited to urban setting. The shortage of supply due to the slowing of the logistics the unavailability of the food materials in the grocery stores led to the stockpiling in the urban and peri-urban households. It was observed from the study that on an average the consumption of packed food such as bread, buns, chips, chocolate, sauce, etc., was initially low due to the unavailability of the products in the shops but after the easing of the lockdown the consumption resumed. The consumption of cereals, pulses, vegetables, and fruits was more as residents resorted to stockpiling and storage facility. In the urban areas, there was a surge in online food and grocery buying. The consumption of dairy products increased in the

Table 9.6 Pre- and post-Covid-19 occupation-wise food consumption (kg per month per HH)

Primary occupation	Cereals		Pulses		Milk and milk products Milk and milk		Vegetables		Fruits		Non-veg (Chicken/fish/meat)	
	Pre	Post	Pre	Post	Pre	Post	Pre	Post	Pre	Post	Pre	Post
Agriculture and allied	41.79	49.79	13.71	25.02	19.86	39.29	19.07	33.93	11.68	18.68	24.68	49.89
NTFP-Forest Dependent	17.00	32	4.00	7	5.00	25	4.00	6	3.00	4	5	8
Government	25.58	42.41	7.25	10.98	15.75	32.58	8.58	17.41	4.63	10.55	22.21	49.54
Private	47.36	70.93	13.79	23.7	21.50	41.93	19.21	34.64	11.96	20.1	26.46	52.68
Daily wage	27.28	42.5	8.33	13.49	14.33	27.44	10.83	20.44	7.75	13.42	19.3	39.86

urban and peri-urban areas, as it is one of the essential commodities. The economic impact on the various income classes is also evident from the study. The per capita consumption and savings have been largely altered for the lower income classes. Many of the lower income classes earnings were hit by an abrupt halt of all the economic activities. Some were forced to look for alternate employment opportunities. The savings were being exhausted at a slow pace. As the informal markets were closed, the acquisition of food materials were difficult for lower income classes but the price of the food materials, such as non-veg items and oil, reduced, whereas for cereals and pulses it continued to be the same. Vegetables, onions, and other exported food products saw a steep rise in prices.

As people had to stay in the house and work from home, elevated energy consumption was inevitable in urban and peri-urban areas. In rural areas, people resorted to going back home and search for income alternatives. Household cooking added to extensive use of LPG, induction, and other forms of cooking fuel in urban and peri-urban areas whereas biomass energy was utilized more in rural areas for cooking; sporadic rise in LPG prices also affected the rural households. An increase in entertainment, online interaction, videos, and other leisure activities are also responsible for increased energy use. Daily household needs in changing season has to be met hence, use of AC, coolers, heaters, laptops, phones, and other electrical appliances have added to the excess and longer period of electricity consumption by households.

Sectoral Policy Insights for Household Emission Reduction

10

India, with more than 1.3 billion population, is the third biggest emitter of greenhouse gases after the USA and China. India has also been in the midst of demographic, economic, and technological transitions. The trends and forecasts promise the nation to become one of the fastest growing consumer appliances markets propelled by rapid rate of urbanization, enhanced household electrification, raising per capita income and women participation in the workforce (Price Waterhouse Coopers. Future of consumer durables and electronics in India—the changing landscape 2018). Factors like growing aspirational middle class with better disposable incomes, and better as well as convenient access to consumer products have also been contributing towards increasing emissions levels, both per capita and per household over time. India made significant progress towards improving energy access and digital inclusion over the last few years. India's annual per capita energy consumption is 0.6 tonnes of oil equivalent, about a third of the world average according to the Economic Survey of 2018–2019 (MoF 2019). Share of renewable energy in the total energy mix has been consistently increasing where solar energy generation is leading the pack. Around 75 crore people in India gained access to electricity between 2000 and 2019 (IEA 2020). According to the Government of India announcement, India had achieved its goal of providing electricity to almost every household in March 2019 (IEA 2020). Indian population is projected to increase to 1.7 billion by 2042 (United Nation 2017) and projected to overtake China as the world's most populous country by 2025. Projection done by NITI Aayog indicates that the electricity consumption for the residential sector is expected to increase 6–13 times by 2047.

As a significant emitter of greenhouse gases, India is expected to play a critical role in global climate change mitigation. However, being a developing country starting from a low emissions base, India needs to balance the perception of an energy-deficient climate deal-breaker with coal is the major raw material for energy generation to a forerunner of a low carbon future with exceptionally ambitious renewable energy generation target. Keeping development agenda in the forefront, together with eradicating poverty and commitment to follow a low carbon path to progress, India's NDC has communicated three numeric targets for 2030. India's goal is to reduce overall emission intensity and improve energy efficiency of its economy over time and at the same time protecting the vulnerable sectors of economy and segments of society. The three targets are to reduce emission intensity by 33–35% from 2005 levels, achieve an installed power capacity of 40% from

S. Nautiyal et al., *Determinants and Stratification of Household Carbon Footprint*,
https://doi.org/10.1007/978-981-96-2739-4_10

non-fossil fuel sources and create an additional carbon sink of 2.5–3.0 billion tons carbon dioxide equivalent from forest and tree cover (INDC document).

According to IEA analysis, India could avoid an additional 15% of annual energy demand and 300 million tonnes of CO_2 emissions during the period 2000–2018 as results of various steps taken. The Government of India has initiated a range of new policy measures to further improve the energy efficiency which are expected to reduce energy intensity over 30% by 2030 and approximately 50% by 2040 as projected by IEA. This projection is based on current and announced policies where coal remains the primary source of electricity generation with a gradual decline in overall share in the energy mix. Coal share is expected to fall to 47% from 74% in 2030. The renewables share is estimated to expand from the current 18–45% in 2040, resulting in 40% carbon-intensity reduction. Despite its falling share of coal in the energy mix, the coal consumption is expected to rise over time, and it is assessed that the power sector emissions would increase from 1.2 Gt CO_2 in 2017 to 2.1 Gt CO_2 by 2040.

There have been increasing research, guidelines, and policy to promote Green Building in India and around the world in more than two decades. India has 10–12% of the world's Green Structures. In India, Green Building Code is an amalgamation of various regulations and guidelines such as State by-laws, National Building Codes, Energy Conservation Building Code and standard rating programme such as Indian Green Building Council, Leadership in Energy and Environmental Design (LEED), Green Rating for Integrated Habitat Assessment (GRIHA). LEED is the globally adopted green building certification guidelines which was developed by US-Green Building Council in 1993 and has been adopted by more than 57 countries till date. The codes have been evolving since then. India had 11.64 million GSM of LEED-certified space in 2014 which has increased to more than 50 million GSM in 2020. That covers 14 lakh houses (6.3 billion sq. ft) with Green Building tag which is constantly increasing (27th November, The Economic Times). India Green Building Council has a mandate to achieve 10 billion sq. ft. of green building footprint by 2022. Report by The Energy Research Institute shows that India can save 8400 Megawatts of power if all homes adhere to the green building standards. Among all green built space in India, residential green space is very low.

There are other such top-down policies aiming at achieving a low carbon progressing economy of India. Policies aligned with people's aspirations and needs should be formulated with detail understanding of types of consumers about who, what, where, and how much. The findings of the study are intended to suggest concern authorities to identify the scope to curtail GHG emissions considering the demand side behaviour and suggest strategies for attaining sustainable development with minimized carbon intensification. The large increase in consumption patterns along the income to be understood for strategizing emission reduction.

Household energy (electricity, LPG, and biomass) being the highest contributing item (39%) in the total average HCF, it requires special attention to pathways for reducing HCF. Household energy in terms of LPG and electricity together result in 1911 $KgCO_2e$ per households per year and 3350 $KgCO_2e$ per household per year for IC-6 in India. More than 80% of the households are dependent on non-clean energy.

More than two-fold jump in electricity consumption from IC-4 (5–10 L) to IC-5 (10–30 L) was found as a critical point to investigate, where the increase was prevalent irrespective of rural or urban households. An increased use of household appliance and energy-intensive heating and cooling are associated with higher energy consumption in higher and highest annual income groups. Energy-climate policy has traditionally focused mainly on production activities, with consumption addressed principally in terms of the energy efficiency of products. While direct emissions from production of energy, other products and services are an integral part of the IPCC mitigation discourse, less emphasis has been given to indirect emissions by advocating lifestyle and end-use consumption changes (Moberg et al. 2019; Dietz et al. 2009; Dubois et al. 2015). Policies and programmes to encourage people to save energy have relied largely on technological interventions and decision-making driven by economics: consumers will save energy to save money (Fredericks 2015). Driven by methodological considerations, supply side and carbon dioxide removal options feature prominently in the emerging pathway literature, while much less attention has been given to the role of demand-side approaches. It confirms that insufficient attention has been paid to demand-side measures, which are found to be inextricably linked to supply-side decarbonization and able to complement supply-side measures (Mundaca et al. 2019). Recognizing the use of energy-related technologies and services, utilities in the US and EU implemented behaviour-focused programmes in the residential sector. These have been found to be as cost-effective and in most of the cases cheaper than technology-oriented Demand Side Management (DSM) measures. Policies to affect these choices focus on price changes (e.g. subsidies for energy-efficient goods) and information disclosure (e.g. mandated energy-use labels on appliances and autos) (Allcott and Mullainathan 2010). The results from behavioural analysis of the current study reveals that the section of people that needs to cut down their household energy use is the most aware section of the society with commendable knowledge on climate change and need for household carbon emission reduction. Then, what do they lack? The barriers understood in this process of adoption are low ease of access to consider adoption of green energy (Solar) or lack of motivation to shift towards energy-efficient appliances. 46% of middle and high-income urban households will consider shifting to solar energy, half of them is ready to invest for the transition, whereas the other half will opt for it, if subsidized.

With 17% of Total Final Consumption (TFC), the transport sector is the third-largest energy consumer and this sector accounted for 26% of total emission in 2017. Energy consumption for transport has more than doubled in 10 years and more than tripled since 2000. While interventions in the electricity sector have largely been driven by strong policy commitments, there is also increasing action in the transport sector. Traditionally, the policies and regulations under the automobile sector have been incremental and gradual; however, recently there has been evidence of adopting progressive approaches in the policy formulations. There has also been increased focus on fuel efficiency standards, emission norms, safety and alternate fuels which are expected to impact overall environment and the society. Similar to household energy, travel CF, both private and public travels have shown steep increase in higher and highest income groups of the study; however, the larger share comes under essential travels. It provides a low scope to reduce travel emission from public travels with respect to consumers' end. With relation to low carbon travel alternatives, respondents of middle and higher income (except IC-6) are assigning high weightage and willingness to low-emission public travels rather than shifting to low-emission private vehicles such as EV or hybrid vehicles. The motivation may be led by connectivity and convenience, as they state.

10.1 Policies Contributing Towards Access Improvement

10.1.1 Energy

Pradhan Mantri Sahaj Bijli Har Ghar Yojana (SAUBHAGYA) programme was launched in 2017 with the aim of providing last mile connectivity and electricity connections to all un-electrified households. Rural Electrification Corporation (REC) has been designated as nodal agency for the Saubhagya scheme. Over 26 million households have been provided access to electricity by this scheme by March 31, 2019. Due to the access to electricity, use of kerosene as a fuel for lamps in non-electrified households is expected to considerably reduce and hence emission attached with kerosene burning.

Pradhan Mantri Ujjwala Yojana (PMUY) was launched in 2016, which aims to safeguard the health of women and children by providing them with a clean cooking fuel LPG. Initially, targeting 50 million LPG connections, the scheme reached over 80 million covering 715 districts in September 2019. IEA projections indicate that the share of the population relying on the use of traditional biomass for cooking will continue to decline, dropping initially to around 500 million, a third of population in 2030, and a quarter by 2040. The potential saving of cumulative emission due to access to cleaner cooking fuel by replacing biomass is enormous. It is estimated that burning 1 kg wood emits 1.9 kg of CO_2.

10.1.2 Transport

The National Electric Mobility Mission Plan (NEMMP) 2020 was conceived and launched in 2013 to promote Electric Vehicles (EV) and Hybrid Electric Vehicles (HEV) to enhance the domestic manufacturing capabilities. The NEMMP envisages strategies to achieve the objective of efficient, environmentally friendly, affordable EVs by 2020. The NEMMP 2020 aims for production of 6–7 million units per annum by 2020 with full range of Electric Vehicles (EVs) and fuel savings of 2.2–2.5 million tons by 2020 (Table 10.1).

In April 2015, the Government launched a scheme for the Faster Adoption and Manufacturing of (HEVs) Electric Vehicles (FAME India). The scheme aims to facilitate the creation of a market for HEVs and EVs across all vehicle segments—cars, motorbikes, buses, commercial vehicles, etc. The promotion of HEVs and EVs

Table 10.1 Sale of electric vehicle in India

Type	FY 2020	FY 2019	FY 2018
Two wheeler	152,000	126,000	54,800
Four wheeler	3400	3600	1200

is expected to ensure much needed energy and ecological security. The FAME India Scheme Phase II is proposed to be implemented over a period of 3 years from 1 April 2019, for faster adoption of electric mobility and growth of electric and hybrid technology.

10.2 Policies Contributing Towards Efficiency Improvement

10.2.1 Energy

Unnat Jeevan by Affordable LEDs and Appliances for All (UJALA) programme was launched in 2015 to provide Light Emitting Diode (LED) bulbs to domestic consumers with a target to replace 770 million incandescent bulbs with LED bulbs by March, 2019. This initiative has a potential of annual estimated GHG reduction of 79 million tons of carbon dioxide. According to Ministry of Power Press release (14 January 2020), over 36.10 crore LED bulbs, 71.61 lakh LED tube lights, and 23.10 lakh energy-efficient fans distributed across the country under UJALA Scheme. This has resulted in estimated energy savings of 47 billion kWh per year with avoided peak demand of 9590 MW, estimated GHG emission reduction of 38 million tons of CO_2 per year and estimated cost saving of INR 18,935 crore per year in electricity bills of consumers. 59.18 lakh Solar Study Lamps distributed to school going students under Solar Study Lamp scheme which is also expected to contribute towards emission reduction in a small way. UJALA programme has radically pushed down the price of LEDs available in the market globally and helped to create local manufacturing facilities. They use 90% of less energy than incandescent bulbs for the same amount of light and half as much as compact fluorescents.

Ministry of Power has issued guidelines to all States to convert all existing consumer meters into smart meters in prepaid mode. Operation of smart meters in prepaid mode would allow consumers to pay as per their own financial convenience and electricity consumption requirements. Energy Efficiency Services Limited (EESL), a JV between CPSUs in the power sector, has been providing smart metering services to various utilities where more than 9 lakh smart meters have been installed in the state of Uttar Pradesh, Haryana, Bihar, NDMC-Delhi, and Andhra Pradesh. This intervention would be critical in managing inefficiencies and tracking usage of appliances in individual homes. Smart meters with real-time displays would keep consumers informed and could influence long-term changes in their behaviour on energy usages. Real-time monitoring through digital coordination platforms offers improvements in asset utilization, improved quality of service, and potentially lower emissions.

The Central Government in consultation with the Bureau of Energy Efficiency (BEE) has notified new energy performance standards for Room Air Conditioner (RACs) on 30th October 2019. 24 °C default setting has been made mandatory from *Jan 1, 2020* for all RAC covered under the star-labelling programme. Additionally, the Indian Seasonal Energy Efficiency Ratio (ISEER) as per the new standards will range from 3.30 to 5.00 for split and (2.70–3.50) for window air conditioners, which will be applicable from first January 2021 onwards. BEE launched the voluntary star-labelling programme for RACs in 2006, and this programme became mandatory on 12th January 2009. Thereafter, in 2015, voluntary star-labelling programme for inverter room air conditioners was launched and which was made mandatory with effect from first January 2018. The BEE star-labelling programme for Room Air Conditioners now covers both fixed and inverter RAC up to a cooling capacity of 10,465 watts (2.97 TR). Continual enhancement in performance levels has resulted in substantial energy efficiency improvement of about 43% in the minimum energy performance standards (MEPS) for split units. According to the available BEE data, electricity savings of approximately 6.09 BU was achieved through sale of star rated air conditioners in 2016–2017 (TERI 2019).

10.2.2 Transport

Bharat Stage (BS) emission standards are laid down by the government to regulate the output of air pollutants from internal combustion engine and spark-ignition engine equipment, including motor vehicles. The central government has mandated that vehicle makers must manufacture, sell, and register only BS-VI (BS6) vehicles from April 1, 2020. Government in January 2016 took a major policy decision to leapfrog directly to the BS-VI Emission Norms from BS-IV for gasoline and diesel vehicles (with 10 ppm Sulphur fuel). The BS VI standard also includes a limit for ammonia emissions, which should not exceed 10 ppm for compression ignition as well as positive ignition engines. The leapfrog to BS-VI norms by 2020 skipping BS-V emission norms will reduce 82% in the emission of particulate matter and 68% in NOx emission in diesel cars.

In 2014, the Government of India finalized the country's first passenger vehicle fuel efficiency standards. The standard set the efficiency targets for new cars at the equivalent of 130 gCO_2/km in 2017 and 113 gCO_2/km in 2022. The standards will keep roughly 50 million tons of CO_2 out of the atmosphere in 2030 alone. The fuel efficiency standard will be based on the average weight of the fleet that manufacturers sell in a given year.

10.3 Policies that can Foster Financial Implications (Incentives, Subsidy, and Taxes)

10.3.1 Energy

The government of India has committed to achieve 100 GW of solar capacity in the country by 2020, of which 40 GW is expected to be achieved through decentralized and rooftop scale solar projects. The MNRE has several policies to incentivize and facilitate rooftop installations. Both the Central Government and State Nodal Agencies (SNAs) offer multiple subsidy schemes to the people for installing rooftop Solar Photovoltaic (SPV) systems. The majority of rooftop SPV projects implemented in India are driven by national and state-specific policies driven by net and gross metering concepts coupled with capital subsidy for rooftop SPV systems. Net metering allows customers who generate their own electricity from solar to feed unused electricity back into the grid and be compensated for that. Subsidies and other schemes offered by the government enable consumers to enhance rooftop solar energy generation. There are loans available at lower interest rates that overcome the problem of high upfront payment associated with solar projects. Electricity Regulatory Commissions, in more than 15 different states, have notified the regulations for grid connected rooftop SPV projects. The state policies mostly encourage self-consumption and sale of surplus power to utility at predefined rates (Table 10.2).

In 2018, India spent a total of USD 25 billion on subsidies for LPG (USD 17B) and gas consumption (USD 4 billion). Kerosene and cooking gas (LPG) are the only oil products subsidized by the government in 2019. As of June 2019, 247 million LPG consumers have joined the PAHAL, an initiative for Direct Benefit Transfer to the consumers started in 2013 (relaunched in 2014). Government data show that households that have received a free connection and the subsidy via the PAHAL programme buy around 3.5 cylinders of LPG per year and are displacing around one-quarter of the fuelwood used in the households that have taken up the scheme, or 6% of total biomass used in India's households (IEA 2018).

10.3.2 Transport

One of the major passenger car taxation reforms happened was in the year 2006. This had led to creation of whole new segment of automobile unique for India—The sub 4 m SUV and Sedan segment, with engine capacity of less than 1200 cc for petrol and 1500 cc for diesel (Table 10.3).

India has adopted Goods and Services Tax (GST) from July 1, 2017; GST is an amalgamating several Central and State taxes into a single tax facilitating a common national market. GST rate for small cars, two/three wheelers and commercial vehicles have been fixed at 28% with additional Cess of 1 and 3% for small petrol and diesel cars, while 15–22% Cess would be applicable on larger cars including luxury cars and SUVs. To promote electric vehicles, the tax rate of the same has been kept considerably lower at 5% than the fossil fuel cars. Under Union Budget 2019–2020,

Table 10.2 Some state-specific incentives for Solar Rooftop PV installation

S. no	State	Incentives	Target
1	Telangana	• 30% subsidy on capital for installation from MNRE • 20% subsidy on installation of RTPV up to 3 kW capacity for domestic sector • 100% of sanctioned load for consumers • Electricity duty exemption	5000 MW by 2020
2	Gujarat	• No limit to residential rooftop capacity • Subsidy of 40% given for up to 3 kW systems • 20% subsidy for systems between 3 and 10 kW	3200 MW from RTPV
3	Karnataka	• 30% capital subsidy • Domestic consumer tariff for projects of 1–10 kW: INR 2.97–3.99/unit • Priority loans up to 10 lakh INR given for individuals	2400 MW from RTPV
4	Kerala	• Net metering • Priority loans provided up to 10 lakhs for individuals • Electricity duty exemption	150 MW from RTPV
5	Rajasthan	• Net metering • For domestic customers, excess above 100 units will be purchased at INR 3.14 kWh • 40% capital subsidy for system size up to 3 kW • 20% subsidy for system size from 3 kW up to 10 kW	30,000 MW by 2024–2025

Table 10.3 Taxes on vehicles in India

Tax type	Vehicle category	Tax/Cess rate (%)
GST	Passenger vehicles (petrol, diesel, CNG, electric hybrid)/ commercial vehicles/three wheelers/two wheelers	28
	Electric vehicles	5
Compensation cess	Passenger vehicles (petrol, CNG, LPG) <4 m in length and <1200 cc engine	1
	Passenger vehicles (diesel) <4 m in length and <1500 cc engine	3
	Mid-size passenger vehicles (>4 m in length with <1501 cc engine)	17
	Large passenger vehicles (>4 m in length with >1500 cc engine)	20
	SUVs (>4 m in length with >>1500 cc engine and >169 mm ground clearance)	22
	Hybrid vehicles (except small)	15
	>350 cc two wheelers	3

the Government announced to provide additional income tax deduction of Rs 1.5 lakh (US$ 2146) on the interest paid on the loans taken to purchase EVs (MoF). The government, in a recent move, has approved green licence plates for electric vehicles in order to encourage people to use them. The purpose behind is their easy identification for proposed benefits such as concessional toll, preferential treatment for parking, and free entry in congested zones.

10.4 Policies and Initiatives to Enhance Consumer Knowledge and Awareness

Standards and Labelling (S&L) programme, instituted by Bureau of Energy Efficiency (BEE) has been facilitating a shift to energy-efficient appliances since 2006 by prescribing minimum standards for efficiency and by providing informative labels to help consumers choose energy-efficient appliances. Energy labels can be used stand alone or complement energy standards. Information in a label enables consumers to compare the energy consumption of similar products and factor lifetime running cost into their purchasing decision. An endorsement label is also provided for some products which provides a “certification” to inform prospective purchasers that the product is highly energy efficient for its category. A star rating, ranging from 1 to 5 in the ascending order of energy efficiency is provided to products registered with the BEE.

As part of its mandate to promote energy efficiency and conservation, BEE has been undertaking multimedia energy awareness campaign to inculcate the habit of conservation and sustainability. The objective is to enhance awareness level among the public on energy efficiency and inform them about the virtues of adopting energy conservation. Print, electronic and social media was used effectively for propagating messages on energy efficiency and energy conservation. People are also made aware of the quality and standard issues of energy-intensive appliances.

BEE initiated Eco Niwas Samhita (Energy Conservation Building Code) prescribes minimum energy performance through energy-efficient envelop design. Eco Niwas Samhita, Part–I for Residential Sector is developed and launched in 2018 to set minimum building envelope performance standards to limit heat gains and heat loss as well as for ensuring adequate natural ventilation and day lighting. Energy Efficiency Label for Residential Buildings launched in 2019. The key objective of the programme is to make a transparent instrument over the energy performance of a home which will gradually lead to an effective model while determining home prices in future. The objective of the labelling programme is to make an energy performance of a home an instrument of comparison while deciding over the home prices in the future. It also aims to provide a benchmark of household energy efficiency standards to create a consumer-driven market transformation solution for energy efficiency in the housing sector.

The government has conceptualized a scheme called Green Urban Transport Scheme (GUTS) for improving and upgrading the public transport in urban areas along the low carbon pathway. The scheme envisages to reduce the carbon footprint and to promote low carbon sustainable public transport system. The scheme would focus on creating non-motorized transport infrastructure and increasing access to public transport. The scheme is expected to promote low carbon public urban transport among urban dwellers and reduce the use of personal vehicles in urban areas.

10.5 Policies that may Improve Energy Efficiency and Lower Emission Further

At present, India does not have a robust national policy on end of life of vehicles. The government had published a draft policy in 2015 for consultation with stakeholders and public. The Voluntary Vehicle Fleet Modernization Programme (V-VFMP) is aimed at encouraging people to get their vehicles scrapped and replace them with modern, more fuel-efficient and less-polluting ones by offering tax and other benefits. The stated policy is yet to be notified. Currently, there are guidelines formulated by the Central Pollution Control Board (CPCB) that are focused on dismantling and recycling of old vehicles in an environmentally sound manner. Effective vehicle scrappage policy will expedite removal of old and inefficient vehicles from the road and incentivize consumers to invest in more efficient mobility options.

In 2017, NITI Aayog has deliberated a plan (Methanol Economy) towards transitioning to alternate fuel methanol produced from coal, biomass. Looking into the potential of operational methanol economy, this can be considered as one of the effective low carbon pathways for emission reduction at household level. Methanol produced from high ash coal, agricultural residue, CO_2 from thermal power plants and natural gas can be used to further reduce GHG emissions in India. Methanol can be blend with petrol, diesel, and LPG. Data shows 15% blending of methanol with gasoline can potentially reduce GHG emission by 20% in terms of particulate matter, NOx and SOx. In 2018, Assam Petro-chemicals has launched canisters-based methanol cooking fuel programme. The programme is an extension the vision of reducing import of crude and an effort to provide clean, cost-effective, and pollution-free cooking medium. Methanol cook stoves can result in a minimum of 20% savings for households in comparison to LPG. After huge success at pilot scale, the methanol cooking programme is being scaled up to 1,00,000 households in the states of Uttar Pradesh, Maharashtra, Gujarat, Telangana, Andhra Pradesh, Goa, Karnataka, Jharkhand, and Manipur.

Formulation of any climate change relevant policy should take care not to compromise with basic food and energy needs of any of the income groups. The basic food-CFs of any of the income group cannot be reduced or altered because of several factors including nutritional, agro-ecological, social, and cultural attributes. The growth of use of packaged and processed food from IC-1 to IC-6, and same for other household consumables and apparels have also shown 2–3.5 times increase. Eco-labelling, carbon pricing of goods and services, other non-durables including healthcare products, home decor, furnishings, gifts, etc., have shown the highest differences in emissions from the poorest group to the richest group

(87.5–1034.98 $KgCO_2e$). This group of items is going to have massive consumers with the growing and aspiring Indian middle class; thereby low carbon items with proper lifecycle assessment for sustainable block chain, eco-labelling should be done through stringent law with efficient monitoring; subsequent carbon pricing and green subsidy should be assigned on final consumer products. The <2% of Indian population which needs to reduce their emission can pay exorbitant tax or price for high emissions to achieve their desired lifestyle and comfort. The responses to the questions on subsidy and incentives for low-carbon alternatives have provided an insight to the potential factors influencing their motivation and willingness. Long-term health impact is one of the concerns which can motivate them to reduce their footprint. Half of that population still believe that climate and environmental concerns are overstated. Innovations in awareness and widespread and intense dialogues among citizens, concern authorities, academia, and urban local bodies may be key in this regard. Structured role of media, both mainstream and social media, through a government institutionalized mechanism will be essential.

Unlike other studies, the findings of the current research reveal a comparatively higher share of CF from wastes in rural India along with its mounting quantity in urban and peri-urban India. The rural waste-CF is also somewhat influenced by the changing attitude of households about manure and wastes from kitchen gardens. The organic concept of circular economy in rural India to be revived with further expansion of those practices in urban and peri-urban India is required for carbon-friendly management of wastes.

In a broader framework of policies, initiatives should be taken to promote and encourage consumption of local produce food, public transport, efficient home appliances, etc. Strategies should be developed to provide better subsidy on capital investment for renewable energy generation at the household level (decentralized implementation approach at district level, village level). Another pertinent step will be to provide policy support to conduct more research of carbon emission impact on processed food, food packaging, recycling, etc., and large-scale consumer awareness activities. Policy to educate, sensitize, and incentivize consumers on both organic and inorganic waste reduction at household level (including e-waste, vehicle, etc.,), most importantly tactical move like "Swachh Bharat" will certainly benefit in emission reduction. Policy to leapfrog implementation of green and sustainable mobility, building and construction, electrical appliances, etc., are other opportunities to consider in the policy domain for reducing household energy and travel HCF. Policy to reduce carbon emission from e-commerce consumptions pattern can help in reducing footprint from the rapidly growing e-commerce purchases. Further research on new pattern and items of consumption such as internet, e-commerce, processed foods, and other non-durables are to be investigated in detail for pathways for emission reduction across the block chains and beyond consumption.

10.6 Relevant for Policy-Makers

- Leveraging the income-consumption relationship to formulate targeted decarbonization and carbonization strategy.
- Encourage more research on emission calculation for a huge number of consumables (where EFs are unavailable), segregating production, and consumption side emissions.
- A framework for consumption data collection, on actual quantity rather than consumption expenditure.
- Policy for encouraging carbon handprint understanding the behavioural aspects of people from different strata.
- It will work as a baseline for spatio-temporal analysis of carbon emission in future.
- A guideline to promote carbon-friendly alternatives acceptable to the society.
- Impact of urbanization and indirect impact of LULC on consumption will be an added value for land use planning.
- Need realized for knowledge and awareness can be used further for the development of awareness programmes.
- The huge set of data is ready to conduct a range of analysis to develop any other understanding with respect to consumption.
- Further investigation and analysis of existing data set has the potential to provide location/region-specific policy for sustainable development.

10.7 Stakeholder Analysis

Primary Stakeholders

- Ministry of Environment, Forest and Climate Change
- Ministry of Statistics and Programme Implementation
- Ministry of Agriculture and Farmers Welfare
- Ministry of New and Renewable Energy
- Ministry of Power
- Ministry of Road transport and Highways
- Ministry of Panchayati Raj
- Ministry of Housing and Urban Affairs
- Department of Consumers Affair
- Nationally Determined Contributions

Secondary Stakeholders

- Localities (U/R/PR)
- Local communities
- NGO's
- Business (Green labelling)
- Other Industries

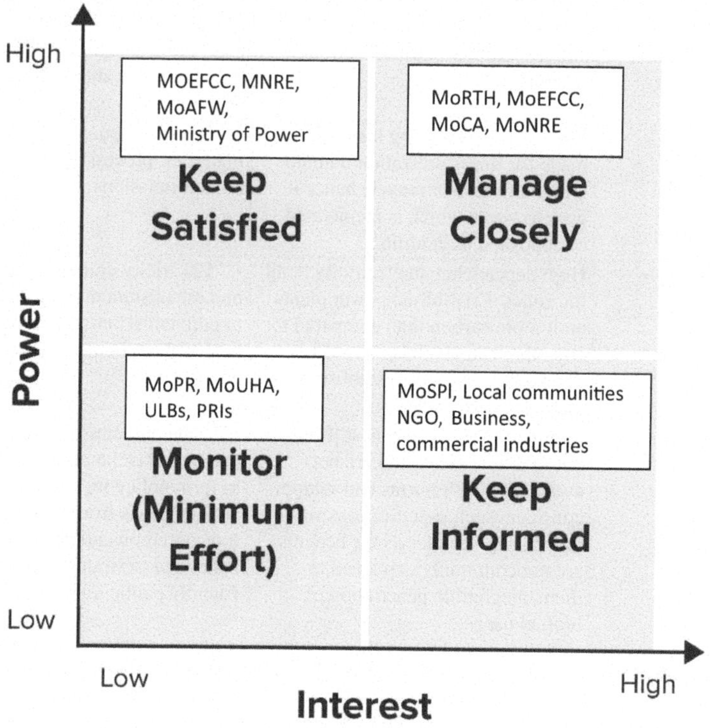

Fig. 10.1 Power/interest grid for stakeholder prioritization

Table 10.4 Applicability of research outcomes for stakeholders

S. no	Actor/agency	Relevant research outcome	Recommendations
1	MoEFCC	Soaring emissions from perpetual use of energy, transport, and food in all the zones. The emissions are bound to increase in the future. A percentage of residents appears incognizant about the climatic impact.	• Carbon emission estimates and relevant policies on carbon trading and carbon pricing. • Awareness programmes. • Policy to determine consumption-based carbon emission (methodology to identify emission hotspots, sectoral emission hotspots, consumer awareness, etc.).
2	MoAFW	Dependency on the use of diesel and electricity-run motors to draw water in the rural areas. Use of pesticides, fertilizers, and use of biomass as fuel are major emitters in agriculture-based households.	• Climate change effect of carbon emissions from cereals, pulses, and policies for encouraging agro-based products for local consumption. • Encouragement for peri-urban agriculture to reduce food miles and local self-sufficiency through optimum use of local resources. • Will have a positive impact on under-consuming peri-urban poor and circular economy.
3	MoNRE	High cost of energy-efficient tools, appliances, solar devices, and non-availability of alternate source of energy.	• Emissions from non-renewable sources for comparative analysis and people's approach towards usage of renewable sources of energy.
4	MoCA	Most of the packaging materials are made for linear utilization with no option of reuse or recycle hence it goes to waste which is incinerated or deposited as landfills.	• Consider action on packaging materials, processed food to reduce emissions.
5	Ministry of Power	High dependency in electricity in all the zones. Coal-based power plants emit more carbon than compared to any other source and coal-fired plants are commonly established in most of the study regions.	• Electricity emission estimates and establishment of infrastructure to curb emissions within the power plant. Increase green energy share.
6	MoRTH	Majority of the residents still uses out-of-life vehicles and are not aware of the BS norms and seldom considers such specifications while purchasing vehicles as the benefits are not commonly explained to them. Negligible penetration of biofuel usage.	• Vehicular emissions and personal travel emission estimates to form policy for vehicle usage. • Emissions from modes of transportation and people's behaviour to strategize people-friendly public transport.

(continued)

Table 10.4 (continued)

S. no	Actor/agency	Relevant research outcome	Recommendations
7	MoPR	The gram panchayats in the rural areas should take up responsibility to disseminate information regarding the carbon emissions due to usage of biomass and appeal to provide alternate sources to meet their requirements.	• Rural emission sources—To encourage social change and service delivery mechanisms for controlled emissions. • Finding sustainable pathways for rural development and economic upliftment of rural households. • Early inculcation of low carbon lifestyles through circular and green economy for the progressing rural middle class. • Promote and incentivize community Forest concept.
8	MoHUA	Increasing trend was observed in direct proportionality between the plot size, land ownership, and emissions.	• Green infrastructure and green building—Laws and incentives. • Sustainable urban planning providing opportunity to urban rich to produce some carbon handprint.
9	NDC	Lower income classes have low per capita emissions when compared to higher income classes.	• Emission information to check requirements for policies and measures to meet the targets. • Strategies for upliftment of economically backward classes and reduce economic inequalities through the benefits of carbon pricing and taxes.
10	Local communities	Particles pertaining to reduction of carbon footprint were not observed. The reduction in footprint from planting trees constitutes less than 10% of their total emissions.	• Information regarding pattern of consumption and expenditure which contributes to carbon emissions and provide reasonable measures to reduce the emissions.
11	NGOs	Minimum knowledge of sustainable farming practices and climatic phenomenon. Need of climate-friendly cooking practices. Lack of local people-centric solution.	• Enabling role towards provoking positive lifestyle change to reduce emissions.
12	Businesses	High willingness of the customers to buy products with green labels and other products which saves energy and is at par or less with non-efficient product's pricing.	• Product information and customers' willingness to pay for sustainable goods and services. • Green labelling, carbon pricing.

(continued)

Table 10.4 (continued)

S. no	Actor/agency	Relevant research outcome	Recommendations
13	Organizations	Information of the percentage of households demand different pricing for energy-efficient products.	• Information about the carbon footprint and product-related emissions which will help them to form internal policies. • Technology access and cost-effective alternatives.

The estimates and qualitative findings can also be used as applied outcomes for local decision-making and which can be scaled up for state-wise and national planning. This can further produce new insights for international climate change knowledge, policy, and practice. As an immediate effort, a formal procedure for reporting back research outcomes to communities, and policy-makers through consultative way can be very effective. These strategies will assist project knowledge to be relevant and accessible for peoples to adopt sustainable lifestyles in the study areas as well as more meaningful use in bottom-up planning. The estimates with segregation across sectors and socioeconomic classes have the potential to contribute towards reviewing existing and formulating new policies on carbon trading, and pricing. The recommendations may include sustainable pathways for rural development and economic upliftment of rural households through local produce and circular economy while looking at pathways for reduction in high emissions in waste management and biofuel use in rural India. Early inculcation of low-carbon lifestyles through circular and green economy principles is essential for the progressing rural middle class, as it can promote sustainable development, reduce environmental impact, and create long-term economic benefits for both individuals and communities. Technology access and cost-effective alternatives, awareness on high emitting pattern of consumerism, and expenditures that contribute to carbon emissions with socially and economically acceptable measures for the community can help households to adopt sustainable lifestyles. Another important aspect emerged as response from the household is information on the carbon footprint and product-specific emissions will help consumers to decide on their consumption rationally (Fig. 10.1; Table 10.4).

10.8 Limitation of the Study

- Unavailability of recent households' consumption data has restricted the comparative analysis of this primary data set prepared from survey to compare with any other meaningful recent secondary data/national statistics.
- Further analysis of more socioeconomic and behavioural variables as determinants of consumption pattern was restricted because of time constraint. Similar

impact of time constraint limited the effort to establish relationship between landscape parameters and consumption behaviour from including in this report.

- The change in consumption (specifically food items) pattern during COVID-19 pandemic could not be captured adequately because of many uncertainties. The main parts of the questionnaire had extensive questions and the additional COVID-19-related questionnaire in the annexure made the survey very exhaustive. This resulted in insufficient and incorrect data collection in a few sections of this additional questionnaire. Further investigation is required to understand actual quantified impact of the pandemic/lockdown on food consumption.
- A few conversion factors are found to be very dynamic (internet) and have very limited references (packaged food items) in Indian context, which may result in <2% variations in HCF estimations.

11 Way Forward

In recent years, India has faced mounting challenges posed by its significant carbon footprint, much of which originates from inefficient waste management practices across rural, urban, and peri-urban areas. The issue underscores the critical need for targeted policy interventions aimed at transforming waste management and consumption practices into drivers of sustainability and resilience. The study proposes a comprehensive policy framework designed to address these challenges through a multifaceted approach.

The transportation sector of India, responsible for approximately 11% of the nation's total carbon dioxide emissions in 2016, notably contributes to this environmental impact (Bansal and Goyal 2020). The rapid expansion of vehicle sales, particularly in the two-wheeler segment, has been a key driver of this escalating emissions profile (Bansal and Goyal 2020). To address this challenge, India has implemented various policies and initiatives focusing on both individual and household measures to mitigate climate change. Factors such as fuel efficiency, comfort, and brand preference significantly influence consumer decisions when purchasing two wheelers (Bansal and Goyal 2020). Acknowledging the transportation sector's impact on emissions, the government has introduced policies to encourage the adoption of more fuel-efficient and environmentally friendly vehicles, including electric two wheelers (Bansal and Goyal 2020). In addition to supply-side interventions, India has also enacted measures to regulate the demand for conventional, high-emission vehicles. One such demand-side policy includes the implementation of fuel economy standards aimed at enhancing vehicle efficiency and reducing their carbon emissions (MoEFCC 2018). Furthermore, India has invested in enhancing public transportation infrastructure, such as expanding metro systems and promoting alternative modes of transport like electric vehicles and non-motorized options (TERI 2019).

A pivotal initiative in promoting electric vehicles is the introduction of the Faster Adoption and Manufacturing of Hybrid and Electric Vehicles (FAME) scheme. This scheme provides financial incentives for purchasing electric vehicles and supports

S. Nautiyal et al., *Determinants and Stratification of Household Carbon Footprint*,
https://doi.org/10.1007/978-981-96-2739-4_11

the development of charging infrastructure (NITI Aayog 2020). Additionally, state governments have implemented their own policies, including subsidies, tax benefits, and exemptions from certain fees, to encourage the adoption of electric vehicles (ICCT 2021). India's National Electric Mobility Mission Plan (NEMMP) 2020 represents a pivotal initiative aimed at accelerating the uptake of electric vehicles. The plan seeks to bolster fuel security by advocating for hybrid and electric vehicles, supported by government incentives and subsidies (MoHI 2012). Additionally, initiatives such as the Smart Cities Mission and the Atal Mission for Rejuvenation and Urban Transformation (AMRUT) prioritize sustainable urban transport solutions, emphasizing non-motorized modes of transportation and reducing reliance on fossil fuel-powered vehicles (MoHUA 2017).

At the individual and household levels, several initiatives have been implemented to promote sustainable practices. The Unnat Jyoti by Affordable LEDs for All (UJALA) scheme has distributed millions of LED bulbs to households, resulting in substantial reductions in electricity consumption and related emissions (Ministry of Power 2021; Kumar et al. 2019). This effort has been recognized for its role in cutting down annual carbon dioxide emissions by millions of tons, highlighting the significant impact of widespread adoption of energy-efficient lighting solutions (IEA 2020). The Pradhan Mantri Ujjwala Yojana has provided clean cooking fuel (LPG) to millions of rural households, replacing traditional biomass stoves and significantly reducing indoor air pollution (Ministry of Petroleum and Natural Gas 2022; Smith et al. 2014). The shift to LPG has not only improved health outcomes but also contributed to decreased greenhouse gas emissions by reducing dependence on wood and other biomass for cooking (World Bank 2021). Furthermore, the Smart Cities Mission promotes urban centres to adopt sustainable solutions such as waste management and public transportation, which have an indirect impact on individual behaviours (Ministry of Housing and Urban Affairs 2015; Joshi and Agrawal 2016). By prioritizing the integration of technology and sustainability in urban planning, the mission has led to the development of more efficient waste collection systems and encouraged non-motorized transport options like cycling and walking (Ghosh 2019).

In India, demand-side regulations for climate change mitigation have mainly concentrated on energy efficiency and conservation. The Perform, Achieve and Trade (PAT) scheme, although primarily aimed at industries, also indirectly affects consumer goods and services (Bureau of Energy Efficiency 2020; Oak and Bansal 2022). By establishing energy efficiency targets for industries, the PAT scheme fosters a market for energy-saving technologies, ultimately benefiting consumers with more efficient appliances and reduced energy costs (Rajagopal et al. 2019). The government has also implemented stricter fuel efficiency standards for vehicles, prompting manufacturers to create more fuel-efficient models and influencing consumer preferences (Ministry of Road Transport and Highways 2018; Bhattacharya and Patel 2019). These standards are part of India's comprehensive strategy to reduce transportation emissions, which also include incentives for electric vehicle adoption under the Faster Adoption and Manufacturing of Hybrid and Electric Vehicles (FAME) scheme (NITI Aayog 2020). In the building sector, the Energy

Conservation Building Code (ECBC) establishes minimum energy performance standards for new commercial buildings, which can affect residential energy-use patterns in mixed-use developments (Bureau of Energy Efficiency 2017; Shetty et al. 2020). The implementation of the ECBC has resulted in significant energy savings in new construction projects and has heightened awareness about the importance of energy-efficient building design (RICS 2018).

Firstly, promoting circular economy practices emerges as a cornerstone of this policy agenda. Scaling up traditional organic waste management practices from rural to urban settings and encouraging household-level initiatives such as composting and recycling can establish a robust circular economy model. To support these efforts, substantial subsidies should be directed towards decentralized renewable energy generation, thereby not only reducing waste but also promoting cleaner energy sources like solar and wind power.

Secondly, enhancing research and awareness on carbon emissions associated with consumption habits is essential. Studies focusing on the carbon footprint of processed foods, food packaging, and recycling processes can provide critical insights. Concurrently, launching extensive awareness campaigns aimed at educating consumers about the environmental impacts of their consumption choices is imperative to foster a culture of sustainability.

Thirdly, implementing waste reduction policies is crucial. This includes targeted measures to reduce both organic and inorganic waste streams, including e-waste and vehicle waste. Initiatives like "Swachh Bharat" should be further empowered to improve waste management practices and mitigate-associated emissions. Additionally, policies supporting green mobility options, energy-efficient buildings, and appliances can significantly curb household energy consumption and related emissions.

In parallel, robust research initiatives must be pursued to underpin these policy actions. This includes developing precise emission calculations for various consumables and establishing frameworks for accurate consumption data collection. Studying the behavioural factors influencing carbon footprints and conducting spatial and temporal analyses of carbon emissions can inform location-specific policies and urban planning efforts. Furthermore, research should explore the indirect effects of urbanization and changes in land use on consumption patterns and carbon emissions, alongside strategies for effectively educating consumers about sustainable consumption practices.

Looking forward, translating research findings into actionable policies at district, village, and urban levels will be crucial for widespread adoption of carbon-friendly practices. Continued investment in research to understand emerging consumption patterns, such as internet use and e-commerce, will also be essential to address evolving environmental impacts. Collaborative efforts between policy-makers, researchers, and community stakeholders will be pivotal in refining policies and promoting sustainable development across India.

In summary, India's comprehensive strategy to tackle emissions from the transportation sector encompasses a range of measures targeting both the supply and demand sides. By promoting electric and fuel-efficient vehicles, expanding public transportation infrastructure, and supporting non-motorized transport initiatives, India is making substantial progress in lowering its carbon emissions and combating climate change. By implementing these comprehensive policy and research recommendations, India has the potential to significantly reduce its household carbon footprint associated with consumption. This concerted effort not only addresses environmental concerns but also paves the way for a more sustainable future, ensuring a healthier and more resilient environment for generations to come. Furthermore, India's holistic strategy to lower emissions and combat climate change at the individual and household levels encompasses a blend of regulatory measures, incentive programs, and public awareness campaigns. Initiatives like the NAPCC, UJALA, and the Smart Cities Mission, along with demand-side regulations such as the PAT scheme and fuel efficiency standards, are driving substantial advancements in promoting sustainable practices and cutting down the country's carbon footprint.

References

Adnan M, Safeer R, Rashid A (2018) Consumption-based carbon footprint analysis approach in urban slum and non-slum areas of Rawalpindi. Habitat Int 73:16–24. https://doi.org/10.1016/j.habitatint.2017.12.012

Ahmed M (2023) The science of climate change. In: The Palgrave handbook of global sustainability. Springer International Publishing, Cham, pp 195–222

Alfredsson EC (2004) "Green" consumption—no solution for climate change. Energy 29(4):513–524. https://doi.org/10.1016/J.ENERGY.2003.10.013

Allcott H, Mullainathan S (2010) Behavior and energy policy. Science 327(5970):1204–1205

Arunan I, Crawford RH (2021) Greenhouse gas emissions associated with food packaging for online food delivery services in Australia. Resour Conserv Recycl 168:105299

Bailis R, Drigo R, Ghilardi A, Masera O (2015) The carbon footprint of traditional woodfuels. Nature Climate Change 5(3):266–272

Baiocchi G, Minx J, Hubacek K (2010) The impact of social factors and consumer behavior on carbon dioxide emissions in the United Kingdom. J Industrial Ecol 14(1):50–72. https://doi.org/10.1111/j.1530-9290.2009.00216.x

Bansal S, Goyal P (2020) Policy measures for reducing carbon emissions in India's transportation sector. J Environ Policy 28(4):577–594

Batchelor CL, Dowdeswell JA, Rignot E, Millan R (2019) Submarine moraines in Southeast Greenland fjords reveal contrasting outlet-glacier behavior since the Last Glacial Maximum. Geophys Res Lett 46(6):3279–3286

Ben Hassen T, El Bilali H, Allahyari MS (2020) Impact of COVID-19 on food behavior and consumption in Qatar. Sustainability 12(17):6973

Bera A, Meena RS, Barman A, Saha P (2024) The use of biochar to reduce carbon footprint: toward net zero emission from agriculture. In: Biochar production for green economy. Academic Press, pp 389–411

Betz J (2014) 6 A shift in India's domestic and international climate policy? In: Environmental and climate change in South and Southeast Asia. Brill, pp 159–184

Bhattacharya S, Patel N (2019) Fuel efficiency norms in India: impact on consumers and manufacturers. J Transp Policy 74:92–100

Bhowmick S (2021) Climate finance: the $100 billion goal and India's climate strategy. Observer Research Foundation

Boguski TK (2010) Life cycle carbon footprint of the National Geographic magazine. The Int J Life Cycle Assess 15:635–643

Bureau of Energy Efficiency (2017) Energy conservation building code

Bureau of Energy Efficiency (2020) Perform, achieve and trade (PAT) scheme

Cao Q, Kang W, Xu S, Sajid MJ, Cao M (2019) Estimation and decomposition analysis of carbon emissions from the entire production cycle for Chinese household consumption. J Environ Manage 247:525–537. https://doi.org/10.1016/j.jenvman.2019.06.044

S. Nautiyal et al., *Determinants and Stratification of Household Carbon Footprint*, https://doi.org/10.1007/978-981-96-2739-4

Census of India (2011) Office of the Registrar General & Census Commissioner. Government of India, New Delhi

Chen S, Wang C, Guo Y, Yang Y, Blondeel M (2023) International and domestic leadership for fulfilling carbon neutrality in emerging economies: comparative evidence from China and India. J Chinese Govern 8(4):533–562

Connolly M, Shan Y, Bruckner B, Li R, Hubacek K (2022) Urban and rural carbon footprints in developing countries. Environ Res Lett 17(8):084005

de Coninck H, Revi A, Babiker M, Bertoldi P, Buckeridge M, Cartwright A., Wollenberg L (2018) Strengthening and implementing the global response

Dietz T, Gardner GT, Gilligan J, Stern PC, Vandenbergh MP (2009) Household actions can provide a behavioral wedge to rapidly reduce US carbon emissions. Proc Natl Acad Sci 106(44):18452–18456

Dubash NK, Khosla R, Kelkar U, Lele S (2018) India and climate change: Evolving ideas and increasing policy engagement. Annu Rev Env Resour 43(1):395–424

Dubash NK et al (2022) India's climate change policy: commitments, challenges and contradictions. Environ Politics 31(5):775–799

Dubois D, Ordabayeva N (2015) 13 Social hierarchy, social Status, and status consumption

Energy Statistics Report (2020) National Statistical Office, Ministry of Statistics and Programme Implementation, Government of India

Fredericks CF (2015) Plenary energy. W Va L Rev 118:789

Ghosh S (2019) Smart Cities Mission and urban sustainability in India. Urban Stud J 56(6):1121–1136

Ghosh A (2023) India's climate diplomacy: balancing national interests and global responsibilities. Brookings India

Gill AR, Viswanathan KK, Hassan S (2018) A test of environmental Kuznets curve (EKC) for carbon emission and potential of renewable energy to reduce greenhouse gases (GHG) in Malaysia. Environ Dev Sustain 20:1103–1114

Gonzalez-García S, Esteve-Llorens X, Moreira MT, Feijoo G (2018) Carbon footprint and nutritional quality of different human dietary choices. Sci Total Environ 644:77–94

Gopalan C (1985) The mother and child in India. Econ Pol Wkly:159–166

Gore C (2015) The post-2015 moment: towards sustainable development goals and a new global development paradigm. J Int Dev 27(6):717–732

Green RF, Joy Edward JM, Francesca H, Sutapa A, Lukasz A, Jon H, Macdiarmid JI, James M, Vetter SH, Pete S, Andy H, Dangour AD (2018) Greenhouse gas emissions and water footprints of typical dietary. Sci Total Environ 643:1411–1418

Hertwich EG, Peters GP (2009) Carbon footprint of nations: a global, trade-linked analysis. Environ Sci Tech 43(16):6414–6420. https://doi.org/10.1021/es803496a

Hirano Y, Ihara T, Yoshida Y (2016) Estimating residential CO2 emissions based on daily activities and consideration of methods to reduce emissions. Build Environ 103:1–8. https://doi.org/10.1016/j.buildenv.2016.02.021

https://www.carbonindependent.org/

https://www.tatapower.com/sustainability/calculate-carbon-footprints.aspx

ICCT (2021) India's electric vehicle policy report

IEA (2018) Global ENERGY & CO2 status report. The latest trends in energy and emissions in 2018

IEA (2019) Energy efficiency 2019

IEA (2020) UJALA scheme: transforming India's lighting market

India GHG Program (2015) India specific air transport emission factors for passenger travel and material transport. India GHG Program Secretariat, Mumbai

IPCC (2007) In: Eggelston S, Buendia L, Miwa K, Ngara T, Tanabe K (eds) IPCC guidelines for national greenhouse gas inventories

Irfany MI, Klasen S (2017) Affluence and emission tradeoffs: evidence from Indonesian households' carbon footprint. Environ Dev Econ 22(5):546–570. https://doi.org/10.1017/S1355770X17000262

Janssen NA, Nyga R, Vanderbeke L, Jacobs C, Ergun M, Buil JB, Verweij PE (2021) Multinational observational cohort study of COVID-19–associated pulmonary aspergillosis. Emerg Infect Dis 27(11):2892

Jemyung L, Oliver T, Keiichiro K (2021) The scale and drivers of carbon footprints in households, cities and regions across India. Glob Environ Chang 66:102205

Jha AK (2023) Climate resilient infrastructure in India: progress and challenges. Econ Pol Wkly 58(15):23–26

Joshi R, Agrawal S (2016) Smart cities mission: challenges and opportunities for urban development in India. J Urban Plan 42(3):45–58

Kaul B, Maldonado B, Michlberger A, Halley S (2023) Analysis of real-world preignition data using neural networks (No. 2023-01-1614). SAE Technical Paper

Koide R, Lettenmeier M, Kojima S, Toivio V, Amellina A, Akenji L (2019) Carbon footprints and consumer lifestyles: an analysis of lifestyle factors and gap analysis by consumer segment in Japan. Sustainability (Switzerland) 11(21):1–25. https://doi.org/10.3390/su11215983

Kumar S, Rathi S, Mukherjee A (2019) Impact assessment of UJALA scheme on energy consumption. Int J Energy Res

Kumar P, Sharma S, Joshi A (2019) The Unnat Jyoti by Affordable LEDs for All (UJALA) scheme: Impacts on electricity consumption and emissions. Energy Policy 130:123–132

Lee J, Taherzadeh O, Kanemoto K (2021) The scale and drivers of carbon footprints in households, cities and regions across India. Glob Environ Change 66:102205

Lenzen M, Pade LL, Munksgaard J (2004) CO2 multipliers in multi-region input-output models. Econ Syst Res 16(4):391–412

Levay PZ, Vanhille J, Goedeme T, Verbist G (2021) The association between the carbon footprint and the socio-economic characteristics of Belgian households. Ecol Econ 186:107065. https://doi.org/10.1016/j.ecolecon.2021.107065

Liu WX, Dou H, Wei ZC, Chang B, Qiu WX, Liu Y, Tao S (2009) Emission characteristics of polycyclic aromatic hydrocarbons from combustion of different residential coals in North China. Sci Total Environ 407(4):1436–1446

Liu X, Song J, Wang H, Wang S (2020) Indirect carbon emissions of urban households in China: patterns, determinants and inequality. J Clean Prod 241(2019):118335. https://doi.org/10.1016/j.jclepro.2019.118335

Long Y, Yoshida Y, Dong L (2017) Exploring the indirect household carbon emissions by source: analysis on 49 Japanese cities. J Clean Prod 167:571–581. https://doi.org/10.1016/j.jclepro.2017.08.159

Loyarte-Lopez E, Barral M, & Morla JC (2020) Methodology for carbon footprint calculation towards sustainable innovation in intangible assets. Sustainability 12(4):1629

Martin R (2021) Rebuilding the economy from the Covid crisis: time to rethink regional studies? Reg Stud Reg Sci 8(1):143–161

Martin A, Markhvida M, Hallegatte S, Walsh B (2020) Socio-economic impacts of COVID-19 on household consumption and poverty. Econ Disast Climate Change 4(3):453–479

Miehe R, Scheumann R, Jones CM, Kammen DM, Finkbeiner M (2016) Regional carbon footprints of households: a German case study. Environ Dev Sustain 18:577–591. https://doi.org/10.1007/s10668-015-9649-7

Ministry of Environment, Forest and Climate Change (2018) India's progress in combating climate change. Government of India

Ministry of Environment, Forest and Climate Change (MoEFCC) (2018) India's fuel economy standards

Ministry of Heavy Industries (MoHI) (2012) National electric mobility mission plan 2020

Ministry of Housing and Urban Affairs (2015) Smart cities mission

Ministry of Housing and Urban Affairs (MoHUA) (2017) Atal mission for rejuvenation and urban transformation (AMRUT)

Ministry of Petroleum and Natural Gas (2022) Pradhan Mantri Ujjwala Yojana: clean cooking for all

Ministry of Power (2021) Unnat Jyoti by affordable LEDs for all (UJALA) scheme

Ministry of Road Transport and Highways (2018) Fuel efficiency norms for vehicles

Moberg E, Walker Andersson M, Sall S, Hansson PA, Roos E (2019) Determining the climate impact of food for use in a climate tax—design of a consistent and transparent model. Int J Life Cycle Assess 24: 1715–1728

MoEFCC (2010) Low carbon lifestyles, centre for environment education. Nehru Foundation for Development (NFD), New Delhi

MoF (2019) Annual report of 2019-20. Ministry of Finance, Government of India, New Delhi

Mukhopadhyay P (2021) Climate adaptation: building resilience through structural transformation. United Nations Conference on Trade and Development (UNCTAD)

Mundaca L, Urge-Vorsatz D, Wilson C (2019) Demand-side approaches for limiting global warming to 1.5 C. Energy Efficiency 12:343–362

Nandy SN (2023) Differential carbon footprint in India–an economic perspective. J Sustain Environ Manag 2(1):74–82

Niamir L, Ivanova O, Filatova T (2020) Economy-wide impacts of behavioral climate change mitigation: linking agent-based and computable general equilibrium models. Environ Model Software 134:104839. https://doi.org/10.1016/j.envsoft.2020.104839

Nicola M, Alsafi Z, Sohrabi C, Kerwan A, Al-Jabir A, Iosifidis C, Agha R (2020) The socio-economic implications of the coronavirus and COVID-19 pandemic: a review. Int J Surg 78:185–193

NITI Aayog (2019) National electric mobility mission plan 2020

NITI Aayog (2020) FAME India scheme

NITI Aayog (2021) National multidimensional poverty index: baseline report. Government of India, New Delhi

Oak H, Bansal S (2022) Enhancing energy efficiency of Indian industries: effectiveness of PAT scheme. Energy Economics 113:106220

Obringer R, Rachunok B, Maia-Silva D, Arbabzadeh M, Nateghi R, & Madani, K (2021) The overlooked environmental footprint of increasing Internet use. Resour Conserv Recycl 167(105389):04

Pachauri R (2008) Climate change and sustainability science. Sustain Sci 3:1–3

Parikh J, Panda M, Ganesh-Kumar A, & Singh V (2009) CO_2 emissions structure of Indian economy. Energy 34(8):1024–1031

Pathak H, Jain N, Bhatia A, Patel J, Aggarwal PK (2010) Carbon footprints of Indian food items. Agr Ecosyst Environ 139(1-2):66–73

Poudel K, Subedi P (2020) Impact of COVID-19 pandemic on socioeconomic and mental health aspects in Nepal. Int J Soc Psychiatry 66(8):748–755. https://doi.org/10.1177/0020764020942247

Prys-Hansen M (2022) Competition and cooperation: India and China in the global climate regime

PWC (2018) PwC reporting perspectives

Rajagopal D, Phadke A (2019) Prioritizing electric miles over electric vehicles will deliver greater benefits at lower cost. Environ Res Lett 14(9):091001

Ramachandra TV (2012) Decentralised carbon footprint analysis for opting climate change mitigation strategies in India. Renew Sustain Energy Rev 16(8):5820–5833

Ramachandra TV, Sreejith K, Bharath HA (2014) Sector-wise assessment of carbon footprint across major cities in India. Assess Carbon Footprint Diff Ind Sect 2:207–267

RICS (2018) Implementation of energy conservation building code in India.

Roy J, Chakravarty D, Dasgupta S, Chakraborty D, Pal S, Ghosh D (2018) Where is the hope? Blending modern urban lifestyle with cultural practices in India. Curr Opin Environ Sustain 31(June 2017):96–103. https://doi.org/10.1016/j.cosust.2018.01.010

Saryal R (2018) Climate change policy of India: Modifying the environment. South Asia Res 38(1):1–19

Saluja MR, Yadav B (2006) Social accounting matrix for India 2003-04. India Development Foundation

Sengupta S (2019) India's engagement in global climate negotiations from Rio to Paris. India in a warming world: Integrating climate change and development, pp 114–141.

Sethi M (2015) Location of greenhouse gases (GHG) emissions from thermal power plants in India along the urban-rural continuum. J Clean Prod 103:586–600

Shetty S, Kishore P, Kini P, Acharya RR, Raj A (2020) Energy Conservation Building Code (ECBC) based optimum wall composition with respect to thermal transmittance and thickness for different commercial pockets of Tier-1 city in temperate climatic zone of India. Procedia Manufacturing 44:229–236

Sikdar C, Mukhopadhyay K (2016) Impact of population on carbon emission: lessons from India. Asia-Pacific Dev J 23(1):105–132

Sinha SK, Subramanian KA, Dutta V (2018) India's progress and plan of action in addressing climate change and its role in setting an effective, cooperative and equitable climate policy architecture. Climate Change Environ Sustain 6(1):1–15

Smith KR, Bruce N, Balakrishnan K et al (2014) Millions dead: how do we know and what does it mean? Methods used in the comparative risk assessment of household air pollution. Annu Rev Public Health 35(1):185–206

Sri P, Banerjee R (2023) Characteristics, temporal trends, and driving factors of household carbon inequality in India. Sustain Prod Consum 35:668–683

TERI (2019) Sustainable urban transport in India: challenges and opportunities

Teske S (2019) Achieving the Paris climate agreement goals: Global and regional 100% renewable energy scenarios with non-energy GHG pathways for+ 1.5 C and+ 2 C. Springer Nature, p 491

UN (2014) Climate change 2014 synthesis report. IPCC, Geneva, pp 1059–1072

UNFCCC (2018) Katowice climate package. United Nations Framework Convention on Climate Change

UNFCCC (2022) COP27 reaches breakthrough agreement on new "loss and damage" fund for vulnerable countries. United Nations Framework Convention on Climate Change

UNFCCC (2023) COP28: key outcomes. United Nations Framework Convention on Climate Change

United Nations (2017) World population prospects 2017: summary of results. United Nations, Department of Economic and Social Affairs, Population Division. https://population.un.org/wpp/

Wiedenhofer D, Lenzen M, Steinberger JK (2011) Spatial and socioeconomic drivers of direct and indirect household energy consumption in Australia. CSIRO Publishing, Collingwood, pp 251–266

WilsonLindsay (2011) Shrink your product footprint. https://shrinkthatfootprint.com/shrink-your-product-footprint/

World Bank (2021) The role of clean cooking in reducing household air pollution and climate change

WRI (2014) Securing rights, combating climate change. How strengthening community forest rights mitigates climate change. https://www.wri.org/research/securing-rights-combating-climate-change

www.3.epa.gov/carbon-footprint-calculator/carbonfoot.com/calculator.aspxcarboninde-pendent.org

Zha J, Wu J, Zhao D (2017) Effects of land use and cover change on the near-surface wind speed over China in the last 30 years. Prog Phys Geogr 41(1):46–67

Zhang J, Zou X, Muhkia A (2023) Reconsidering India's climate diplomacy and domestic preferences with a two-level approach. Int J Climate Change Strat Manag 15(5):671–689

The manufacturer's authorised representative in the EU is Springer Nature Customer Service Centre GmbH, Europaplatz 3, 69115 Heidelberg, Germany. If you have any concerns regarding our products, please contact ProductSafety@springernature.com

Printed and bound by CPI Group (UK) Ltd, Croydon, CR0 4YY
01/10/2026
02225944-0010